Cellulose - Fundamentals and Conversion Into Biofuel and Useful Chemicals

Edited by Rajesh Banu Jeyakumar, Kavitha Sankarapandian and Yukesh Kannah Ravi

Published in London, United Kingdom

Cellulose - Fundamentals and Conversion Into Biofuel and Useful Chemicals
http://dx.doi.org/10.5772/intechopen.104273
Edited by Rajesh Banu Jeyakumar, Kavitha Sankarapandian and Yukesh Kannah Ravi

Contributors
Hassan Sabbaghi, Rebika Baruah, Archana Moni Das, Ridha Lafi, Hajer Chemingui, Imed Montasser, Amor Hafiane, Ibtissem Ounifi, Fatma Khaled, Malak Kahloul, Hanan Ajari, Umut Yucel, Eda Ceren Ceren Kaya, Lili Fu, Bing Wang, Ke Zhang, Quanxing Zheng, Zhongya Guo, Qi Zhang

First published in London, United Kingdom, 2023 by IntechOpen
IntechOpen is the global imprint of INTECHOPEN LIMITED, registered in England and Wales, registration number: 11086078, 5 Princes Gate Court, London, SW7 2QJ, United Kingdom

British Library Cataloguing-in-Publication Data
A catalogue record for this book is available from the British Library

Additional hard and PDF copies can be obtained from orders@intechopen.com

Cellulose - Fundamentals and Conversion Into Biofuel and Useful Chemicals
Edited by Rajesh Banu Jeyakumar, Kavitha Sankarapandian and Yukesh Kannah Ravi
p. cm.
Print ISBN 978-1-80355-216-3
Online ISBN 978-1-80355-217-0
eBook (PDF) ISBN 978-1-80355-218-7

Meet the editors

Dr. Rajesh Banu Jeyakumar is an Associate Professor in the Department of Biotechnology, Central University of Tamil Nadu (CUTN), Thiruvarur, India. He is also a Visiting Professor at the Center for Excellence in Environmental Studies, King Abdul Aziz University, Saudi Arabia. He has 18 years of teaching experience at national and international universities. He is a visiting researcher at East China Normal University, China, and a former post-doctoral fellow at Sungkyunkwan University (SKKU), South Korea. He is the author of 350 publications in indexed journals. He is a senior editor of *e-Prime – Advances in Electrical Engineering, Electronics and Energy*. He has authored and edited five books and published fifty chapters and fourteen conference proceedings. He obtained seven major grants from the Government of India and two consultancies from industries. He guided 17 Ph.D. students and 125 postgraduate engineering students in environmental engineering and science. Dr. Jeyakumar is listed among the world's top 2% of scientists by Stanford University, USA. He received the 2018 Think of Ecology award from Hiyoshi, Japan. His research centers on converting waste to energy, value-added product recovery from waste, biological wastewater treatment, nutrient removal, membrane bioreactors, and microbial fuel cells.

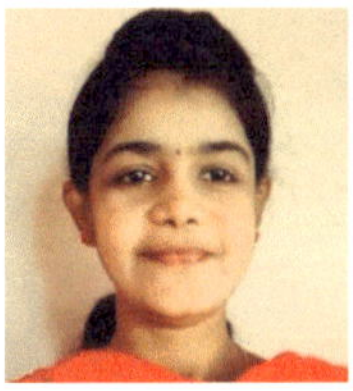

Dr. Kavitha Sankarapandian is an Assistant Professor in the Department of Biotechnology, Karpagam Academy of Higher Education, Coimbatore, Tamilnadu, India. She completed her Ph.D. in the Department of Civil Engineering at Anna University Chennai, India. She received her Bachelor\'s degree in Microbiology and Master\'s in Biotechnology at Manonmaniam Sundaranar University, Tirunelveli, India. Her major research includes energy generation from biomass, micro and macroalgae biorefinery, lignin valorization, and microbial fuel cells. She has authored about 90 articles in SCI and SCIE journals with a cumulative impact factor of 750. Her publications achieved 4000 citations. She has one authored book, two edited books, and 14 book chapters. In addition to this, she is a potential reviewer and editorial board member in various international journals. She has received three best poster awards at International conferences. She was ranked among the Top 2% of Scientists worldwide by Stanford University, USA, in 2020.

Dr. Yukesh Kannah Ravi is an Assistant Professor at the Center for Organic and Nanohybrid Electronics, Silesian University of Technology, Gliwice, Poland. He received a bachelor's degree in Civil Engineering, a master's degree in Environmental Engineering, and a Ph.D. in Environmental Engineering from Anna University Chennai, Tamilnadu, India. His research focuses on converting various waste biomass to bioenergy/biofuels, biological wastewater treatment, biomass biorefinery, microbial electrochemical cells, and value-added product recovery. He has authored/co-authored fifty-five peer-reviewed scientific journals and twenty book chapters. He is also a reviewer and editorial board member for various international journals. Dr. Ravi has received three best poster awards at international conferences.

Contents

Preface

Recently, many researchers have focused on identifying a cost-effective approach to generate biofuel, chemicals, and value-added products from various biomass. Generating alternative fuels from renewable feedstock helps with waste management, produces low-cost fuel, and reduces negative environmental effects. To achieve this, a broader knowledge of selecting renewable feedstock is essential. From an economic and environmental point of view, cellulose is considered a renewable, versatile, and biodegradable material. It is the most abundantly available biopolymer in different biomass such as plants, bacteria, tunicates, and algae. This book provides basic information about the various applications of cellulose conversion into biofuels and value-added products. It presents the basics of cellulose conversion, an integral part of the biorefinery concept, and discusses the possible routes of biofuel/value-added products derived from cellulose.

Rajesh Banu Jeyakumar
Department of Biotechnology,
Central University of Tamil Nadu,
Thiruvarur, Tamil Nadu, India

Kavitha Sankarapandian
Department of Biotechnology,
Karpagam Academy of Higher Education,
Coimbatore, Tamilnadu, India

Yukesh Kannah Ravi
Center for Organic and Nanohybrid Electronics,
Silesian University of Technology,
Gliwice, Poland

Section 1

Environmental Application

Chapter 1

The Coffee Residues and the Esparto Fibers as a Lignocellulosic Material for Removal of Dyes from Wastewater by Adsorption

Ridha Lafi, Hajer Chemingui, Imed Montasser and Amor Hafiane

Abstract

Biosorption onto lignocellulosic products such as coffee residues and esparto fibers in natural and modified forms have been identified as a potential alternative to the existing biosorbents applied for dye removal from wastewater. The efficiency of each material has been discussed with respect to the operating conditions and the chemical modifications. The investigated thermodynamics and kinetics studies were exposed also in terms of equilibrium isotherms and fitted kinetic models. Moreover, the crucial role of the chemical structures of the cellulosic fibers as an affecting factor on the mechanism of the adsorption process was evaluated and compared. The different treatment methods showed an improvement in terms of removal and maximum adsorption capacity. In fact, in some cases the removal capacity can be increased to 99% and the maximum adsorption capacity can reach 67 mg/g. On the other hand, the different investigations showed that the study data fitted to the known model such as Langmuir isotherm and pseudo-second-order kinetic.

Keywords: dye, biosorption, lignocellulosic products, adsorption isotherms, kinetics

1. Introduction

Water pollution due to the rejected industrial hazard has become global issue of concern. Many industries such as textile industry use dyes to color their products and thus produce wastewater containing organics with a strong color. Dyes are used in different industries such printing, textile and cosmetic. Therefore, an important quantity of these dyes is rejected in the environment, thus causing a real ecotoxicity for human and animal health [1, 2]. Dyes are reported toxic, non-degradable, and harmful to the environment and human and animal health [3, 4]. Generally, based on their solubility and chemical characteristics, the dyes are categorized into acid, basic, direct, mordant, vat, reactive, disperse, azo and sulfur dyes [5, 6]. Among the most common methods for the removal of dyes from wastewaters include adsorption, chemical precipitation, coagulation-flocculation, ion exchange, membrane technology, and electrochemical methods [7]. Adsorption is one of the most commonly used techniques for the removal of dyes from wastewater [4]. While other methods limit their applications based on the high cost of materials used, researchers have turned their attention to develop easily available, low-cost, renewable, green, and efficient adsorbents such as agricultural wastes [8–11]. The coffee residue and the esparto fiber were the typical lignocellulosic wastes and by-products, respectively. The coffee residue can be obtained during the treatment of raw coffee powder with hot water or steam for the instant coffee preparation. Every year an important quantity of coffee can be collected and different wastes can be rejected during the preparation of coffee powders from coffee beams which can cause environment pollution [12–19]. So, collection of these wastes was investigated by different researchers for their use as adsorbent materials in the wastewater treatment. On the other hand, Esparto grass (*Stipa Tenacissima* L) is a tussock grass widely distributed in semi-arid and arid regions in North Africa and southern Spain. The leaves are cylindrical, tough and very tenacious reaching up to 1.5 m in height. Traditionally, esparto has been employed for crafts. Humankind has used natural lignocellulosic materials for an amount of applications in daily life [20]. These wastes contain some organic compounds, such as cellulose, hemicellulose, lignin, and waxes. Recently research has been carried out on the application of coffee waste and esparto grass as an adsorbent in water treatment using the adsorption technology to provide references for future researches on the recycling and utilization of these materials in water treatment. This paper reviewed and analyzed the effect of coffee waste and esparto fiber on removal dye from wastewater. In addition, the pretreatment method of coffee waste and esparto fiber was discussed to solve the problem of limited adsorption efficiency. Moreover, this current chapter compares the adsorptive capacity of various forms of coffee wastes and esparto fiber. The most favorable conditions for the decontamination process for each hazardous dye were also discussed. Finally, the existing knowledge on coffee waste and esparto fiber as sorbent was analyzed to provide new directions for further applications of lignocellulosic materials in water treatment.

2. Characterization of the coffee waste and esparto fiber

The interests in utilization of agricultural wastes have been significantly increased and many attempts have been made regarding the use of lignocellulosic materials (either natural substances or agro-industrial wastes and by-products) as economic and eco-friendly options. Agricultural wastes such as lignin, cellulose and hemicelluloses are characterized by high molecular weights. The major chemical constituents of this lignocellulosic biomass include cellulose; hemicellulose, lignin, and their percentage dry weight composition are approximately 39, 31, and 18 wt%, respectively [1]. Cellulose is a linear polymer of β-D-glucopyranose sugar units. The average chain has a degree of polymerization about 9000–10,000 units. Hemicelluloses are amorphous polysaccharide polymers with a low degree of polymerization compared to cellulose. Lignin is a heterogeneous, complex and large molecular structure with cross-linked three-dimensional phenyl-propane polymer of phenolic monomers. Lignins have a highly branched structure, amorphous and made by cross-linking phenolic precursors. The chemical composition of the two lignocellulosic materials is presented in **Table 1**.

		Coffee residue	Esparto fiber
Elemental analysis (wt%)	C	65.8	44.3
	H	28.4	6.5
	N	—	0.6
	O	—	46.5
	Others	5.7	2.1
Cellulose		8.6 (glucose)	45.3
Hemicellulose		36.7	23.7
Lignin		—	23.9
Protein		13	—
Lipids		14	—
Carbohydrates		26	—
Water		45	—
Ash		2	2.1

Table 1.
Chemical composition of lignocellulosic materials.

2.1. Composition and functional groups

The type of functional groups and chemical components in lignocellulosic wastes and by-products are similar but in different amounts [21]. They play an important role in dyes sorption. Lignocellulosic based cellulose hemicellulose and lignin contain different functional groups such as hydroxyl, ether, and carbonyl [22, 23]. The cellulose in the esparto fiber form long chains (or elemental fibrils) linked together by hydrogen bonds and van der Waals interactions Cellulose is usually presented as a major crystalline fraction associated to a minor amorphous. Lafi et al. [24] found that the composition of esparto fiber was 45.3% cellulose, 23.7% hemicelluloses, 23.9% lignin, and 2.1% ash. According to Boehm method [25], the functional groups at the surface of esparto were carboxylic 0.58 mmol/g and basic 0.4 mmol/g, followed by phenolic 0.96 mmol/g and lactonic 0.03 mmol/g. However, the composition of coffee residues was 45% water, 13% protein, 14% lipids, 26% carbohydrates, and 2% ash [23]. The functional groups at the surface of coffee residue were carboxylic 0.225 mmol/g and basic 0.1 mmol/g, followed by phenolic 0.27 mmol/g and lactonic 0.015 mmol/g. These results indicate that the differentiation of the surface chemistry of coffee residues and esparto fibers is directly related to the surface treatment before the adsorption process (**Table 2**).

2.2. FTIR and SEM

FTIR technique has been used to identify functional groups before and after modification of lignocellulosic materials (**Table 3**). These functional groups are responsible for the removal of dyes from water. For example, after crystal violet (CV) biosorption by Esparto fiber [24], the infrared bands 3334, 1721, 1653, 1431, 1381, and 1161 had shifted to 3344, 1737, 1666, 1437, 1369, and 1166 cm^{-1}, respectively, indicating a chemical interaction was occurred between CV molecules and the carboxylate

Adsorbant	pH_{PZC}	S_{BET} (m^2/g)	Boehm titration		
			Carboxylic	Lactonic	Phenolic
Esparto fiber	6.3	20.7	0.58	0.03	0.96
Coffee residue	5.3	2.9	0.225	0.015	0.27

Table 2.
BET and pH_{PZC} parameters of various coffee adsorbents.

Adsorbant	Wavenumber (cm^{-1})	Assignment (functional groups)	Ref.
Coffee residue	3446	Bonded O—H	
	2924	C—H	
	2852	CH_2	
	1728	C═O stretching vibration	
	1658	COO	
	1535	N—H	[26]
	1454	C═O	
	1382	COO	
	1165	C—O—C stretching vibration	
	1029	C═O stretching vibration	
Esparto fiber	3700–3000	free O—H, O—H stretch, and inter-chains H-bonds	
	2918	Asymmetrical stretch vibration —CH_3	
	2844	Symmetrical stretch vibration\—CH_2	
	1721	C═O stretching vibration	
	1658	COO	[24]
	1514	N—H amine/amide groups	
	1246	Symmetric stretching COO	
	1165	C—O—C vibration	
	1033	Stretching vibration of C—O—H	

Table 3.
Interactions of functional groups of coffee waste and esparto fiber and their shifts in FTIR spectra.

and the hydroxylate anions. This result suggests that carboxyl and hydroxyl functions are predominant contributors in dye uptake. Lafi et al. [26] concluded that after the toluidine blue (TB) adsorption by coffee residues, the bands shifted from 3446, 1728, 1658, and 1029 cm^{-1} to 3444, 1737, 1629, and 1031 cm^{-1}, respectively, due to TB adsorption. These indicates that the corresponding functions (—O—H, C═O, COO, C—O and C—O, respectively) are involved on the mechanism of dye adsorption. Based on another study, Lafi et al. [27] studied the comparison of FTIR of coffee residue (CR) and CR modified with cationic surfactants such as cetylpyridinium chloride (CPC) and cetyltrimethylammonium bromide (CTAB). The result shows that with CPC-CR the intensity of —CH_3 (2929 and 1368 cm^{-1}) increased while the peak of —CH_2 (2853 cm^{-1}) was slightly resolved due to the increase in the aliphatic carbon content in CPC-CR. The others conclude that CPC molecules exist on the surface. The

FTIR shows that the band from —CH_2 of CTAB (at 2853 and 1465 cm^{-1}) becomes stronger while the band from —NH_2 and —OH (at 3414 cm^{-1}) is broadened. In this investigation, two new peaks at 1607 cm^{-1} and 1520 cm^{-1} (assigned to aromatic skeletal vibrations and —N=N— stretching vibrations, respectively) appear after adsorption of MR into MCRs. This result is probably related to the new quaternary ammonium group introduced in the adsorbent material after modification. The peaks at 1039 and 1195 cm^{-1} attributed to S=O stretching observed after adsorption of MR into MCR_s indicate that the —SO_3^- groups of MR are involved in the adsorbent material [28]. The FTIR of the modification of extracted cellulose from *Stipa tenacissima* with cetyltrimethyl ammonium bromide (CTAB) showed that the quaternary ammonium group of CTAB was introduced at the surface of modified extracted cellulose (MEC). In fact, the vibration peak from —CH_2 of CTAB (at 2908 and 1427 cm^{-1}) became stronger while the band at 3313 cm^{-1} (from —NH_2 and —OH) is broadened. The new quaternary ammonium group is responsible for the electrostatic interactions between the MEC and the methyl orange (MO) during the adsorption process [29]. Due to their various treatment methods (washing etc.), the adsorbent based coffee residues and esparto fibers present different morphologies before adsorption process. **Figure 1**(left) showed the structure of coffee residue indicating that a porous and homogenous structure with deep pores exists. The surface was not smooth, but scraggy with a variety of cavities involves a small surface area [28]. Regarding the esparto fiber, the others found cellular and irregular textures, related to the existence of heterogeneous layer surfaces with large number of pores characterized by various sizes at the surface [29].

3. Treatment methods of coffee waste and the esparto fiber

After the material is collected, simple pretreatment is required, including washing and drying. The purpose of washing is to remove impurities that can affect the adsorption process [26, 27]. Coffee waste and esparto fiber have the following advantages: (i) porous or various cavity structures, (ii) cellulose, hemicellulose, and lignin structures with abundant functional groups (e.g., —C=O, —COO^-, —COOH and —OH). These functional groups can be involved in adsorption process after simple pretreatment. Lafi et al. [26] used coffee waste after washing and drying to remove CV and TB and obtained adsorption amounts of 125 and 142.5 mg/g, respectively. On the other hand, esparto grass fibers were used to remove CV and TB [24]. The investigation showed that the adsorption capacity was more important with coffee waste (CW), and this result was attributed to the higher lignocellulosic amount in CW. It can be seen that unmodified material have also the potential to adsorb dyes. The treatment methods of coffee waste and the esparto fiber include physical, acid alkali, and others treatment. The physical treatment changes the particle size and surface area of the adsorbent to increase the removal rate [30]. Delil et al. [31] used ultrasonic technology to treat SCG. The particle size was reduced and the surface area was increased from 3.58 to 1.13 m^2/g. In addition, the Zeta potential of this treated SCG became more negative, which enhanced the adsorption of Cd (II). Nabais et al. [32] concluded that esparto fibers have an interesting potential for the production of activated carbons using carbon dioxide as activation agent. The acid treatment leads to structural changes in coffee waste. These changes are accompanied by an increase in the number of acid groups such as the carboxyl group or an introduction of other functional groups to improve its adsorption capacity [33–35]. Ahsan et al. [35]

Figure 1.
SEM of Coffee waste (left) and Esparto fiber (right).

modified spent coffee grounds (SCG) using sulfuric acid as a sulfonating agent. The introduction of sulfonic acid as polar functional groups made the adsorbent surface electronegative with this modification, the adsorption capacity of methylene blue (MB), tetracycline (TC), and chromium (VI) reached 812, 462, and 302 mg/g, respectively. In another study, Raffas et al. [36] prepared activated carbons by the pyrolysis of coffee grounds impregnated by phosphoric acid at 450°C with different impregnation ratios: 30, 60, 120, and 180 wt.%. The lower impregnation ratios (<120 wt.%) led to the best microporous and acidic activated carbons whereas higher impregnation ratios (>120 wt.%) yielded to mesoporous carbons with specific surface areas as high as 925 m^2/g, pore volume as large as 0.7 cm^3/g, and neutral surface. The adsorption uptake of Nylosan Red (N-2RBL: anionic dye of diameter ≈ 2 nm) onto this carbon is 1.75-fold higher than that of a commercial activated carbon ($S_{BET} \approx 1400$ m^2/g). Sodium hydroxide (NaOH) and potassium hydroxide (KOH) are the commonly used reagents for alkali modification that can increase the surface area, pore volume, and some functional groups of coffee waste [37]. An alkali treatment using NaOH was used to modify coffee husk (NaOH-CFCB) [38]. This investigation showed functional groups at the surface of adsorbent and an increase of surface area and pore volume. In another study, Lafi et al. [39] used potassium acetate to activate coffee waste to prepare activated carbon coffee waste (ACCW). The potassium acetate modification created functional groups on the surface of the adsorbent, and also increased the surface area and pore volume, which showed an imported adsorption performance for CR. Several researchers applied cationic surfactant, magnetite nanoparticles, or other chemicals (such as cetyltrimethyl ammonium bromide (CTAB), cetylpyridinium chloride (CPC), dodecyltrimethyl ammonium bromide (DTAB), and N, N-dimethyldodecylamine N-oxide (DDAO) [28, 27] to modify coffee wastes for removing dyes. Cationic surfactant can enhance the hydrophilicity of the surface of coffee wastes, and alter the surface charge characteristics. Khataee et al. [40] synthesized a new adsorbent based Fe_3O_4-loaded waste coffee (Fe_3O_4-CHC) by precipitation method. Using this adsorbent, the decolorization rate of acid red 17 (AR 17) decreased from 100% to 74% with the increase of the initial dye concentration. Methylene blue (MB), methyl orange (MO) and rhodamine B (Rh B) were removed from water using CG4 adsorbent abtained from catalyzed waste coffee grounds by $FeCl_3$ [41]. The removal rates of MB, MO, and Rh B by CG4 were 93.8%, 92.9%, and 94.1%, respectively, after ten cycles. The adsorbent prepared by using a magnetic treatment, the adsorbent can be also prepared. However, the continuous adsorption–desorption cycle reduces the adsorption efficiency under acid condition by ion leaching [42]. Moreover, the use of large quantities of magnetic adsorbents can also lead to serious environmental problems [43].

The acid treatment introduces some functional groups and increases the porosity and the surface area. These modifications improve the adsorption capacity. However, the strong acid reagents used are expensive and corrosive, which limit their industrial application. The alkali treatment can transform coffee waste materials to adsorbents with high adsorption capacity. However, the appropriate ratio must be used when the base solution is used.

4. Adsorption of dyes

Most of dyes are not degradable and can cause carcinogenicity, mutagenesis in humans, and can affect the photosynthesis of aquatic organisms [44, 45]. Therefore, it is necessary to treat toxic dyes in wastewater. The following investigations describe the use of coffee waste and esparto fiber to adsorb dyes form wastewater.

Coffee husk was activated using H_3PO_4 to remove MB [46]. Under these acidic conditions, the negative charges on the surface of the adsorbent attract cationic MB contrary to alkaline conditions that increase the repulsive forces. Therefore, the MB removal rate and the maximum adsorption capacity were about 96.9% and 6.82 mg/g, respectively. Tran et al. [47] also used coffee husk to remove MB. It was found that the activated carbon at 108°C followed by KOH treatment had a high adsorption capacity for MB. The adsorption reached equilibrium rapidly (<30 min) and the removal rate was about 99.28% at pH 7.8 at an initial dye concentration of 200 mg/L. The experimental data showed that the maximum adsorption capacity of AH on MB was 418.78 mg/g. This value is very close to the theoretical data (418.78 mg/g) obtained by the Langmuir model. In term of adsorption kinetic, the experimental date fitted to the pseudo-second-order kinetics to describe the adsorption process. Coffee ground powder (CGP) was used as adsorbents to remove Rh B and Rh 6G [48]. In this study, the chemical structure of dye played a crucial role on the adsorption process. In fact, the adsorption capacity of Rh 6G (17.37 μmol/g) was higher than that of Rh B (5.26 μmol/g). This result was related to the dissociated carboxyl group (—COOH) of Rh B that play a repulsion role between adsorbent and dye, and to the dissociated ester group (—$COOCH_3$) of Rh 6G that was responsible to the hydrophobic interactions between the adsorbent and the dye. The adsorbent based on CGW/PPy composite was prepared using the pyrrole polymerization method. The coffee grounds were the raw material and the potassium persulfate was the oxidant [49]. The CGW/PPy composite was used to remove Rh B. The result showed that when pH (6) > pHPZC (3.2), the electrostatic interaction between CGW/PPy and Rh B increased the adsorption capacity; however under a constant pH of 9, the adsorption capacity increased by 1.7 times and the maximum adsorption capacity was about 50.59 mg/g. In term of adsorption equilibrium, the experimental data fitted to the Langmuir and Redlich-Peterson models. Regarding the adsorption kinetic, the experimental data fitted to the pseudo-second-order model. Coffee grounds (CG) were investigated as adsorbent to remove MG [50]. The FTIR analysis showed different functional groups and the SEM morphology revealed cavities with different sizes which provided channels for the adsorption of MG. The study showed that the removal rate of MG increased with the increase in contact time and adsorbent dosage; however, the increase of the initial MG concentration from 50 to 250 mg/L implies a decrease of the removal capacity from 99.63% to 98.69%. In term of adsorption isotherm and adsorption kinetic, the experimental data fitted to the Langmuir and to the pseudo-second-order models, respectively. Concentrated sulfuric acid-activated coffee husk (ACH) as an adsorbent to remove MG was investigated by Murthy et al. [51]. Using an adsorbent dosage of 0.5 g/L, the removal rate of MG reached 90% and the maximum adsorption capacity was about 263 mg/g. Regarding the influence of the pH; when the $pH > pH_{PZC}$ (5.4), the negative charge on the ACH surface attracts MG, which increases the removal rate and reaches the maximum at pH 6.8. Finally, in this it was found the experimental data fitted to the Sips isotherm model and the pseudo-second order kinetic model. SCG as adsorbent was used to remove CR dye [52]. The results showed that removal rate increased with the increase of adsorbent dosage and contact time until reaching equilibrium. Regarding the adsorption process, it was found that the removal rate decreased with the increase of the initial dye concentration. The optimization of the experimental data was performed using RSM, and the results showed that the removal rate was about 89.17% when the adsorbent dosage was 1.87 g, the initial concentration was 48.18 mg/L, and the contact time was 57.96 min. Potassium acetate was used to modify coffee waste [39]. The obtained activated carbon (ACCW) was used as adsorbent to remove CR. After 120 min the adsorption equilibrium was reached. Furthermore, the authors found that the acid condition lead to the protonation of the oxygen-containing functional groups (—OH and —COOH). Under this condition, the positive charged adsorbent surface attract the R-SO_3^- functional groups of the dissociated CR, and the maximum adsorption capacity of CR was about

90.90 mg/g. The authors found that the experimental data fitted to the Langmuir isotherm model and the pseudo-second-order kinetic model. Coffee husks (WCH) were used as adsorbent to remove CR [53]. The result showed that the adsorption efficiency was about 96% at 25°C when the dye concentration was about 12.24 mg/L and the pH was about 4. In term of adsorption isotherm and kinetic, the authors found that the experimental data fitted to the pseudo-second-order model and both the Langmuir model and the Freundlich model. Untreated SCG was used to adsorb CV [54]. The authors indicated that when the SCG concentration was about 2 g/L, the removal rate was about 90%. When the pH was above 5.3, the surface of the adsorbent acquired a negative charge (pH_{PZC} = 5.3) to attract the cationic CV dye by electrostatic interaction. The FTIR analyze showed the appearance of functional groups such as hydroxyl and carbonyl that could be responsible for the adsorption of CV. The experimental data fitted to the Langmuir isotherm model and the pseudo-second-order model. Regarding the thermodynamic parameters, the data showed the adsorption process was exothermic and spontaneous. Pagalan et al. [55] used KOH activation of SCG to adsorb aniline yellow dye (AYD). The SEM indicated that the KOH-modified SCG formed mesoporous and microporous structures, which were favorable for AYD removal by activated carbon. The factors affecting the adsorption were optimized using response surface methodology (RSM) methodology. The AYD removal rate reached 88.72% and the adsorption capacity was about 2.58 mg/g, when the initial AYD concentration was 35 mg/L. The obtained experimental data fitted to the Freundlich and the pseudo-first-order kinetic models. KOH-activated SCG as adsorbent was investigated to remove orange G [56]. Using this adsorbent, the adsorption capacity reached 100 mg/g at 45°C.The experimental data fitted to the Langmuir and Redlich-Peterson models in term of adsorption equilibrium. In terms of adsorption kinetic the experimental data fitted to the pseudo-second-order model and the diffusion with intraparticles was not the unique rate-limiting step. CW was modified by cetyltrimethyl ammonium bromide (CTAB) and cetyl pyridine chloride (CPC) cationic surfactant [27]. Using these adsorbents, the maximum adsorption capacities of MO by CTAB–CW and CPC–CW were 58.82 and 62.5 mg/g, respectively. In term of adsorption kinetic, the experimental data fitted to the pseudo-second-order kinetic model. In another study, Lafi et al. [28] used cationic surfactant, Dodecyltrimethyl ammonium bromide (DTAB) and a zwitterionic surfactant, N, N-dimethyldodecylamine N-oxide (DDAO) to modify coffee residue (CW) to increase affinity for MO anionic dyes. The maximum adsorption capacities of MR using DTAB-CR and DDAO-CR were 76.22 and 66.22 mg/g, respectively. The pseudo-second-order kinetic model fitted to the experimental data. Lafi et al. [24] studied the performance of Esparto grass fibers (EGF) as adsorbent to remove TB and CV from aqueous solutions. Under optimum conditions (25°C, pH 7.0, contact time of 150 min, and adsorbent dose of 2 g/L), the adsorption capacity was about 40.00 mg/g for TB and 43.47 mg/g for CV, and the two equilibrium data were fitted to the Langmuir isotherm. Adsorption of MO dye onto modified extracted cellulose using cationic surfactants was investigated by Lafi et al. [29]. The investigation showed that the maximum adsorption capacity (16.95 mg/g) was obtained under pH 3.7 with 4 g/L of adsorbent at 25°C. Experimental data showed better agreement with the pseudo-second-order kinetic model and Langmuir adsorption isotherm model. **Table 4** presents the optimal adsorption conditions and maximum adsorption capacities of different studies using coffee wastes and esparto fibers to remove dyes from wastewater [57–65].

Adsorbent	Adsorbate	Process variables of adsorption					Adsorption isotherm	Adsorption kinetic	q_{max}	Ref.
		m (g/L)	C (mg/L)	Contact time	pH	T (°C)				
Coffee ground powder	RhB	1	7.18	3 h	2	19	L	PSO2;PSO1	5.26	[48]
Coffee ground powder	Rh6G	1	6.65	3 h	2	19	L	PSO2;PSO1	17.37	[48]
Coffee wastes	CV	5	80–400	20 min	6	20	L	PSO2	125	[26]
Coffee wastes	TB	5	80–400	20 min	6	20	L	PSO2	142.5	[26]
Coffee waste CTAB	MO	2	20–120	4 h	3.5	25	L	PSO2	58.82	[27]
Coffee wastes-CPC	MO	2	20–120	4 h	3.5	25	L	PSO2	62.5	[27]
Coffee residues AC at 400°C	MR	0.05	5–50	60 min	10	25	L-F	PSO2	—	[57]
Coffee residues AC at 500°C	MR	0.05	5–50	60 min	10	25	L-F	PSO2	—	[57]
Coffee residues AC at 600°C	MR	0.05	5–50	60 min	10	25	L-F	PSO2	—	[57]
Magnetic coffee silverskin	MB	0.005	25–1000	6 h	8	20	L	PSO2	556	[58]
Magnetic green coffee	MB	0.5	50	30–300 min	5.5	20	—	PSO2	66.2	[59]
Magnetic coffee silverskin	MB	0.5	50	30–300 min	5.5	20	—	PSO2	99	[59]
Magnetic spent coffee grounds	MB	0.5	50	30–300 min	5.5	20	—	PSO2	78.1	[59]
Coffee residues	MB	20	10–100	18 min	10	30	L	—	4.68	[60]
Coffee residues	MB	0.5	50	24 h	7.5	25	L	PSO2	112	[61]
Pyrolized coffee residues	MB	0.5	50	24 h	7.5	25	L	PSO2	132	[61]
Coffee residues	MB	7	5–60	180 min	—	27	F	PSO2	6.69	[62]
Coffee husks	BY 3G-P	2	15–550	—	2	20	L	PSO2	24.04	[63]
Coffee wastes	RR 3BS	0–15	0–1000	24 h	2–12	25	L–F;L	PSO2; PSO1	179–241	[64]
	RBRN	0–15	0–1000	24 h	2–12	25	—	—	—	[64]
	RR 3BS	0–15	0–1000	24 h	2–12	25	—	—	—	[64]
Coffee husks	FG	1	10–100	100 min	2	—	L–F	PSO1	96.6	[65]
Coffee husks	MB	1.6	20	50 min	5	30	L	PSO2	6.82	[46]

Adsorbent	Adsorbate	Process variables of adsorption					Adsorption isotherm	Adsorption kinetic	q_{max}	Ref.
		m (g/L)	C (mg/L)	Contact time	pH	T (°C)				
Coffee husk	MB	1.2	—	—	—	—	F	PSO2	78	[47]
Coffee husk waste	MB	—	500	720 min	7.8	—	L	PSO2	418.78	[47]
Coffee grounds waste	RhB	—	200	—	9	20	L, R–P	—	50.59	[49]
Spent coffee grounds	MG	0.001	50–250	90 min	—	30	L	PSO2	—	[50]
Coffee husks	MG	0.2	25–150	120 min	6.8	50	S	PSO2	264.81	[51]
Spent coffee grounds	CR	2	50	60 min	5	37	L	PSO2	—	[52]
Coffee waste	CR	4	—	120 min	3	25	L	PSO2	90.90	[39]
Coffee husks	CV	—	12.24	—	4	25	L–F	PSO2	—	[53]
Spent coffee grounds	CV	2	100	40 min	6	—	L	PSO2	63.3	[54]
Spent coffee grounds	AYD	0.6	35	150 min	—	25	L	PSO2	2.58	[55]
Coffee grounds	OG	1	20–200	250 min	6	25	L	PSO2	72	[56]
Coffee residue -DDAO	MR	2	—	240 min	3.5	25	L	PSO2	76.22	[22]
Coffee residue -DTAB	MR	2	—	240 min	3.5	25	L	PSO2	66.66	[22]
Esparto grass fiber	CV	2	20–100	150 min	7	25	L	PSO2	43.47	[24]
Esparto grass fiber	TB	2	20–100	150 min	7	25	L	PSO2	40	[25]
Modified extracted cellulose	MR	4	20–100	4 h	3.7	25	L	PSO2	16.95	[29]

Table 4
Optimal adsorption conditions, maximum adsorption capacities and removal rates for dyes removal using coffee wastes and esparto fibers.

5. Adsorption isotherm and kinetic

5.1. Adsorption isotherms

The determination of the adsorption isotherms allows an understanding of the interaction between dyes and the adsorbents. Different adsorption isotherm models (**Table 5**) were used to describe the adsorption process such as Langmuir, Freundlich, Temkin, Redlich-Peterson, and Dubinin-Radushkevich [4–6, 8, 66–70]. Langmuir and Freundlich are the most widely used as adsorption isotherm models. These models are only applicable where sufficient time is provided to allow equilibrium between the

Isotherm	Non-linear form	Ref.
Langmuir	$q_e = \frac{q_m b C_e}{(1+bC_e)}$	[67]
Freundlich	$q_e = K_F C_e^{1/n}$	[68]
Temkin	$q_e = Bln(AC_e)$	[69]
D–R	$q_e = q_m exp(-\beta\varepsilon^2)$	[70]

Table 5.
Lists of adsorption isotherms mostly discussed in the present study.

dye in solution and the dye adsorbed on the adsorbent. During the adsorption process, the dye are expected to be in contact with the adsorption sites and thus retained on the adsorbent surface [5]. Due to the different sources and processing conditions of coffee waste and the different nature of various dyes, the adsorption process could be different. Murthy et al. [71] describes the adsorption process of MB removal using coffee husks by the Freundlich model, which suggests that the adsorption is heterogeneous and multi-layered. However, the Langmuir model describes the adsorption process of MB removal using coffee husks waste, which can describe that the adsorption process follows monolayer adsorption with a uniform distribution of active sites at the surface of adsorbent [47]. In the process of adsorbing CV, both the Langmuir and Freundlich isotherm models fitted to the experimental data of CV removal using coffee husks as adsorbent. However, other investigations describe that the Langmuir model fitted more to the experimental data regarding the CV removal [53, 54].

5.2. Adsorption kinetic

The kinetic model gives information about the rate of the adsorption process, and can also explain the mechanism involved in the removal of dye. Several kinetic models have been applied to the adsorption process, such as pseudo-first-order kinetic, pseudo-second-order kinetic, intraparticle diffusion model, and Elovich model [72–76]. In the case of coffee waste, pseudo-second kinetic describe the adsorption process for the removal of dyes. The adsorption by coffee waste is mainly physiosorption, which is the main limiting factor affecting the adsorption rate of the whole adsorption process. In the adsorption processes using modified coffee waste to remove MO, pseudo-second-order kinetic model describes the process, indicating that physiosorption and hydrophobic interactions are involved [27]. In the case of OG, the quasi-second-order kinetic model describes well the adsorption process. On another note, the results of the diffusion model showed that the adsorption of OG by modified coffee grounds was also controlled by intraparticle diffusion [56]. In the case of

Kinetics model	Equation form	Ref
Pseudo-first-order	$log\,(q_e - q_t) = log\,(q_e) - (\frac{k_1}{2.303})t$	[73]
Pseudo-second-order	$\frac{t}{q_t} = \frac{1}{k_2 q_e^2} + \frac{t}{q_e}$	[74]
Elovich	$q_t = \frac{1}{\beta}\,ln\,(\alpha\beta) + \frac{1}{\beta}\,ln\,(t)$	[75]
Intra-particle diffusion	$q_t = k_{id} t^{1/2+C}$	[76]

Table 6
Lists of kinetic equations.

adsorption of Fast green dye using coffee husks, the process can be described by a pseudo-first-order kinetic model [65]. **Table 6** presents a list of equations used to determine the kinetic behavior of various dyes from wastewater by using coffee wastes and esparto fibers.

6. Mechanism of adsorption

The main constituents of lignocellulosic materials contain various functional groups that play an essential role in dyes adsorption. The mechanism of adsorption relies on the adsorbent characteristics (e.g., surface area, porosity, and surface functional groups) [23]. Molecular properties of dyes (e.g., molecular size, aliphatic vs. aromatic and hydrophobicity), solution properties (pH, ionic strength, and temperature), and the interactions between functional groups of adsorbent and dyes are often involved in the adsorption mechanism [77]. The surface charges of adsorbent and adsorbate id related to the pH, and it is one of the important factors that control the adsorption process. The adsorption regulation could be achieved by measuring the point of zero charges (pH_{PZC}) of sorbents using the zeta potential [78]. At $pH < pH_{PZC}$, protonation of functional groups leads to a positively charged adsorbent that can successfully remove the anionic sorbates. Kyzas et al. [79] reported that the determination of pH_{PZC} played an important role to explain the possible pH-mechanism regarding the adsorption of reactive dye (RB) onto UCR. The pH_{pzc} was about 3.2–3.4 and negatively charged surface of UCR was occurred when the pH values above of 3; however, positively charged occurred at lower pH values. Therefore, at pH < 3.5, the adsorption process involves electrostatic interaction between UCR^+ and SO_3^- of dyes, and at pH > 3.5. The interactions decrease because the dye is still negatively charged, illustrating the reduction of dye (RB) removal. In the case of basic dye (BB), the molecule presented constant positive charge; the UCR was protonated in acidic pH values. Therefore, the adsorption process is very limited. However, increasing the pH of the solution (pH > 3.5) the surface of UCR is charged negatively. In this case, the deprotonation of the surface of UCR is taken place, and transformed to the negatively charged form which provides electrostatic interactions favorable for adsorbing cationic species. In alkaline conditions, the increase of pH solution would convert more groups to $—O^-$ and $—COO^-$, providing electrostatic interactions for cationic species removal. Based on these findings, the more significant adsorption mechanisms of dyes onto coffee residue is through electrostatic interaction. If the $pH > pK_a$, the sorbate surface charge is negative; however, at $pH < pK_a$, the sorbate exists in a positively charged surface. Thus, the adsorption capacity of the adsorbent can be altered by changing the pH solution. Reffas et al. [36] investigated the adsorption of methylene blue and Nylosan Red N-2RBL onto activated carbons prepared by the pyrolysis of coffee grounds. In the case of MB, the mechanism can be explained on the basis of an electrostatic interaction between the ionic dye molecule and the charged carbon substrate. At pH 6, the CGAC30, CGAC60 and CGAC120 activated carbons are negatively charged ($pH > pH_{PZC} \approx 3.7$) while the CGAC180 and CAC are positively charged ($pH < pH_{PZC}$). Therefore, electrostatic repulsion between the MB cation (at pH 6) and CGAC180 (or CAC) is not in favor of adsorption.

In another study Murthy et al. [51] reported that the adsorption of MG by ACH was carried out by van der Waals force and electrostatic interaction. It was also controlled by membrane and intra-particle diffusion. The adsorption was accelerated with the increase of temperature and concentration. Furthermore, Shen and Gondal [48] examined the adsorption mechanism of Rh B and Rh 6G using CGP as adsorbents. The results involve electrostatic interaction and molecular interaction and the adsorption capacity decreased to 0.9 μmol/g for Rh B and 5.5 μmol/g for Rh 6G after 5 cycles. Based on FTIR analysis, Lafi et al. [43] found that the mechanism of adsorption could be explained by the hydrogen bonding and the electrostatic interaction between CR and oxygen-containing functional groups on the activated carbon surface. Cheruiyot et al. [53] provide that the electron sharing or exchange between WCH and CV indicates that the adsorption process is controlled by chemisorption. In another study [27], through the FTIR characterization and mechanism analysis, it was indicated that the process of adsorption of MO may involve hydrophobic–hydrophobic interaction and electrostatic interaction. Based on the FTIR characterization and the proposed adsorption mechanism, Lafi et al. [28] concluded that zwitterionic surfactant (DDAO) is the most efficient for the adsorption of MR onto coffee residues (CR). The mechanism involves different types of interaction such as hydrogen bonds, electrostatic and hydrophobic interactions. Lafi et al. [35] also investigated the adsorption capacity of esparto grass fibers (EGF) for TB and CV removal. Infrared spectroscopy demonstrated that several functional groups were involved in CV and TB binding on EGF such as ester, hydroxyl and amino groups. Binding seemed to be more related to chemisorption with hydrogen atoms of non-ionized carboxyl groups. Sorption behavior of modified extracted cellulose (MEC) from *Stipa Tenacissima* L by cetyltrimethyl ammonium bromide was investigated by Lafi et al. [39]. In this case of MO adsorption on the MEC, the mechanism could be related to hydrophobic interaction and electrostatic interaction.

Therefore, isotherms and kinetics models, along with thermodynamic parameters and activation energy, can clarify the physisorption or chemisorption characteristics of an adsorption process.

Different methods of dyes adsorption on coffee waste and esparto fiber adsorbents or and the adsorption mechanisms are schematically presented in **Figure 2**.

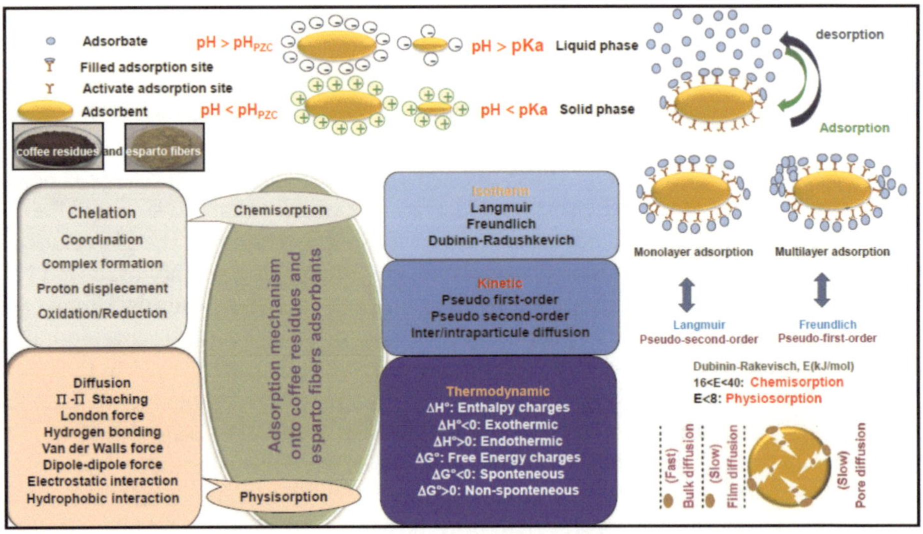

Figure 2.
Schematic illustration for the adsorption mechanism of dyes onto coffee waste and esparto fiber based materials.

7. Conclusions

This chapter was devoted to the use of lignocellulosic material based coffee residues and esparto fibers in the adsorption of dyes from wastewater. This Lignocellulosic material was investigated in natural and modified forms, and showed a potential use as adsorbents in the detoxification of water using the adsorption process. These adsorbents were characterized using different analytical methods such as FTIR, SEM, and BET. In term of kinetic and thermodynamic, this chapter demonstrated that the different experimental data fitted to the known models. We believe that Lignocellulosic material could be used in industrial water purification by removing undesirable chemicals, biological contaminants, and gases from water.

Conflict of interest

The authors declare no conflict of interest.

Abbreviations

AC	activated carbon
AH	activated the hydrothermal carbon
ACH	acid-activated coffee husk
AR17	acid red 17
AYD	aniline yellow dye
ACCW	activated carbon coffee waste
BC	biochar
BET	Bruner-Emmett-Teller
BPA	bisphenol A
BY	Brilliant yellow 3G-P
CH	coffee husk
CR	congo red
CV	crystal violet
CW	coffee waste
CR	coffee residue
CGW	coffee ground waste
CGP	coffee ground powder
CTAB	cetyl trimethyl ammonium bromide
CPC	cetyl pyridine chloride
CGAC	coffee grounds activated carbon
CAC	commercial activated carbon
DTAB	dodecyltrimethyl ammonium bromide
DDAO	N, N-Dimethyldodecylamine N-oxide
ECW	exhausted coffee waste
EGF	esparto grass fiber
FG	Fast green
FTIR	Fourier transform infrared spectroscopy
Fe_3O_4-CHC	Fe_3O_4-loaded coffee waste hydrochar
HC	hydrochar

HAC	coffee husk
KOH	potassium hydroxide
ICO	International Coffee Organization
MB	methylene blue
MO	methyl orange
MR	methyl red
MG	malachite green
MEC	modified extracted cellulose
MCW	Fe_3O_4 loaded magnetic coffee waste
MCR_s	modified coffee residue
NaOH	sodium hydroxide
N-2RBL	Nylosan Red
NaOH-CFCB	NaOH-modified coffee husk
NaOH-SCG	NaOH-modified waste coffee grounds
OG	orange G
pH_{PZC}	pH the point of zero charge
PBD	Plackett-Burman design PPy polypyrrole
PSO1	pseudo-first-order
PSO2	pseudo-second-order
Rh B	rhodamine B
Rh 6G	rhodamine 6G
RH	rice husk
RSM	response surface methodology
RB	Remazol Brilliant Blue RN
RR	Remazol Red 3BS
RB	reactive dye
BB	basic dye
SAC	spent coffee
SCG	spent coffee ground
SEM	scanning electron microscope
SCGB	spent coffee grounds derived biochar
SMCW	surface modified coffee waste
SCBC	spent coffee grounds biochar
TC	tetracycline
TB	toluidine blue
UCR	untreated coffee residue
UACW	ultrasonic assisted coffee waste
WCH	waste coffee husk

Author details

Ridha Lafi[1]*, Hajer Chemingui[1], Imed Montasser[2] and Amor Hafiane[1]

1 Laboratory of Water, Membranes and Environmental Biotechnology, CERTE, Soliman, Tunisia

2 INRAP, LMU, Technopark of Sidi Thabet, Tunisia

*Address all correspondence to: ridha.lafi@yahoo.fr

References

[1] Keskin B, Ersahin ME, Ozgun H, Koyuncu I. Pilot and full-scale applications of membrane processes for textile wastewater treatment: A critical review. Journal of Water Process Engineering. 2021;**42**:102172

[2] Sultana M, Rownok MH, Sabrin M, Rahaman MH, Nur Alam SM. A review on experimental chemically modified activated carbon to enhance dye and heavy metals adsorption. Cleaner engineering and technology. 2022;**6**: 100382

[3] Drumond Chequer FM, de Oliveira GAR, Anastacio Ferraz ER, Carvalho J, Boldrin Zanoni MV, de Oliveir DP. Textile dyes: Dyeing process and environmental impact. In: Eco-Friendly Textile Dyeing and Finishing. 2013

[4] Yagub MT, Sen TK, Afroze S, Ang HM. Dye and its removal from aqueous solution by adsorption: A review. Advances in Colloid and Interface Science. 2014;**209**:172-184

[5] Demirbas A. Agricultural based activated carbons for the removal of dyes from aqueous solutions: A review. Journal of Hazardous Materials. 2009; **167**:1-9

[6] Salleh MAM, Mahmoud DK, Karim WAWA, Idris A. Cationic and anionic dye adsorption by agricultural solid wastes: A comprehensive review. Desalination. 2011;**280**:1-13

[7] RobinsonT GMM, Marchant R, PN. Remediation of dyes in textile effuent: A critical review on current treatment technologies with a proposed alternative. Bioresource Technology. 2001;**21**: 211-222

[8] Solangi NH, Kumar J, Mazari SA, Ahmed S, Fatima N, Mubarak NM. Development of fruit waste derived bio-adsorbents for wastewater treatment: A review. Journal of Hazardous Materials. 2021;**416**:125848

[9] Anastopoulos I, Kyzas GZ. Agricultural peels for dye adsorption: A review of recent literature. Journal of Molecular Liquids. 2014;**200**:381-389

[10] Rangabhashiyam S, Anu N, Selvaraju N. Sequestration of dye from textile industry wastewater using agricultural waste products as adsorbents. Journal of Environmental Chemical Engineering. 2013;**1**:629-641

[11] Gupta VK, Suhas. Application of low-cost adsorbents for dye removal—A review. Journal of Environmental Management. 2009;**90**:2313-2342

[12] Rodrigues da Silva M, Sanchez Bragagnolo F, Lajarim Carneiro R, de Oliveira Carvalho Pereira I, Aquino Ribeiro JA, Martins Rodrigues C, et al. Metabolite characterization of fifteen by-products of the coffee production chain: From farm to factory. Food Chemistry. 2022;**369**:130753

[13] Anastopoulos I, Karamesouti M, Mitropoulos AC, Kyzas GZ. A review for coffee adsorbents. Journal of Molecular Liquids. 2017;**229**:555-565

[14] Torres Castillo NE, Ochoa Sierra JS, Oyervides-Muñoz MA, Sosa-Hernández JE, Iqbal HMN, Parra-Saldívar R, et al. Exploring the potential of coffee husk as caffeine bio-adsorbent—A mini-review. Case Studies in Chemical and Environmental Engineering. 2021;**3**:100070

[15] Mussatto SI, Carneiro LM, Silva JPA, Roberto IC, Teixeira JA. A study on chemical constituents and sugars extraction from spent coffee grounds. Carbohydrate Polymers. 2011;**83**:368-374

[16] Sarasini F, Tirillò J, Zuorro A, Maffei G, Lavecchia R, Puglia D, et al. Recycling coffee silverskin in sustainable composites based on a poly(butylene adipate-co-terephthalate)/poly (3-hydroxybutyrate-co-3-hydroxyvalerate) matrix. Industrial Crops and Products. 2018;**118**:311-320

[17] Del Pozo C, Bartrolí J, Alier S, Puy N, Fàbregas E. Production of antioxidants and other value-added compounds from coffee silverskin via pyrolysis under a biorefinery approach. Waste Management. 2020;**109**:19-27

[18] Setter C, Borges FA, Cardoso CR, Mendes RF, Oliveira TJP. Energy quality of pellets produced from coffee residue: Characterization of the products obtained via slow pyrolysis. Industrial Crops and Products. 2020;**154**:112731

[19] Hoseini M, Cocco S, Casucci C, Cardelli V, Corti G. Coffee by-products derived resources. A review. Biomass and Bioenergy. 2021;**148**:106009

[20] Cerd A. The effect of patchy distribution of Stipa tenacissima L. on runoff and erosion. Journal of Arid Environments. 1997;**36**:37-51

[21] Rangabhashiyam S, Balasubramanian P. The potential of lignocellulosic biomass precursors for biochar production: Performance, mechanism and wastewater application —A review. Industrial Crops and Products. 2019;**128**:405-423

[22] Abdolali A, Guo WS, Ngo HH, Chen SS, Nguyen NC, Tung KL. Typical lignocellulosic wastes and by-products for biosorption process in water and wastewater treatment: A critical review. Bioresource Technology. 2014;**160**:57-66

[23] Zhou Y, Zhang L, Cheng Z. Removal of organic pollutants from aqueous solution using agricultural wastes: A review. Journal of Molecular Liquids. 2015;**212**:739-762

[24] Lafi R, Hamdi N, Hafiane A. Study of the performance of Esparto grass fibers as adsorbent of dyes from aqueous solutions. Desalination and Water Treatment. 2014;**2014**:1-14

[25] Boehm HP. Chemical identification of surface groups. Advances in Catalysis. 1966;**16**:179-274

[26] Lafi R, Fradj A, Hafiane A, Hameed BH. Coffee waste as potential adsorbent for the removal of basic dyes from aqueous solution. Korean Journal Chemistry Engeenering. 2014;**31**:1-9

[27] Lafi R, Hafiane A. Removal of methyl orange (MO) from aqueous solution using cationic surfactants modified coffee waste (MCWs). Journal of the Taiwan Institute of Chemical Engineers. 2016;**58**:424-433

[28] Lafi R, Abdellaoui L, Montasser I, Mabrouk W, Hafiane A. The effect of head group of surfactant on the adsorption of methyl red onto modified coffee residues. Journal of Molecular Structure. 2022

[29] Lafi R, Abdellaoui L, Montasser I, Hafiane A. Removal of methyl orange from aqueous solution onto modified extracted cellulose from Stipa Tenacissima L. International Journal of Environmental Analytical Chemistry. 2020;**1**:1-17

[30] Yadav S, Yadav A, Bagotia N, Sharma AK, Kumar S. Adsorptive potential of modified plant-based adsorbents for sequestration of dyes and heavy metals from wastewater—A review. Journal of Water Process Engineering. 2021;**42**:102148

[31] Delil AD, Olcay G, Gören N. Optimization of adsorption for the removal of cadmium from aqueous solution using Turkish coffee grounds. International Journal of Environmental Research. 2019;**13**:861-878

[32] Nabais JMV, Laginhas C, Carrott MMLR, Carrott PJM, Amorós JEC, Gisbert AVN. Surface and porous characterisation of activated carbons made from a novel biomass precursor, the esparto grass. Applied Surface Science. 2013;**265**:919-924

[33] Yeung PT, Chung PY, Tsang HC, Cheuk-On Tang J, Yin-Ming Cheng G, Gambari R, et al. Preparation and characterization of bio-safe activated charcoal derived from coffee waste residue and its application for removal of lead and copper ions. RSC Advances. 2014;**4**:38839-38847

[34] Cerino-Córdova PJ, Díaz-Flores PE, Soto-Regalado E, Gómez-González R, Bustamante-Alcaántara E. Biosorption of Cu (II) and Pb (II) from aqueous solutions by chemically modified spent coffee grains. International Journal Environmontal Science Technology. 2013;**10**:611-622

[35] Ahsan A, Jabbari V, Islam T, Kim H, Hernandez-viezcas JA, Lin Y, et al. Green synthesis of a highly efficient biosorbent for organic, pharmaceutical, and heavy metal pollutants removal: Engineering surface chemistry of polymeric biomass of spent coffee waste. Journal of Water Process Engineering. 2018;**25**:309-319

[36] Reffas A, Bernardet V, David B, Reinert L, Lehocine MB, Dubois M, et al. Carbons prepared from coffee grounds by H_3PO_4 activation: Characterization and adsorption of methylene blue and Nylosan Red N-2RBL. Journal of Hazardous Materials. 2010;**175**: 779-788

[37] Laksaci H, Khelifi A, Trari M, Addoun A. Synthesis and characterization of microporous activated carbon from coffee grounds using potassium hydroxides. Journal of Cleaner Production. 2017;**147**:254-262

[38] Quyen V, Pham TH, Kim J, Thanh DM, Thang PQ, Van Le Q, et al. Biosorbent derived from coffee husk for efficient removal of toxic heavy metals from wastewater. Chemosphere. 2021; **284**:131312

[39] Lafi R. Adsorption of congo red dye from aqueous solutions by prepared activated carbon with oxygen-containing functional groups and its regeneration. Adsorption Science & Technology. 2018; **0**:1-22

[40] Khataee A, Kayan B, Kalderis D, Karimi A, Akay S. Ultrasound-assisted removal of Acid Red 17 using nanosized Fe_3O_4-loaded coffee waste hydrochar. Ultrasonics Sonochemistry. 2017;**35**: 72-80

[41] Wen X, Liu H, Zhang L, Zhang J, Fu C, Shi X, et al. Large-scale converting waste coffee grounds into functional carbon materials as high-efficient adsorbent for organic dyes. Bioresource Technology. 2019;**272**:92-98

[42] Zhang A, Li X, Xing J, Xu G. Adsorption of potentially toxic elements in water by modified biochar: A review. Journal of Environmental Chemical Engineering. 2020;**8**:104196

[43] Tee GT, Gok XY, Yong WF. Adsorption of pollutants in wastewater via biosorbents, nanoparticles and magnetic biosorbents: A review. Environmental Research. 2022;**212**: 113248

[44] Garg A, Chopra L. Dye waste: A significant environmental hazard. Materials Today: Proceedings. 2022;**48**: 1310-1315

[45] Methneni N, Morales-González JA, Jaziri A, Ben MH, Fernandez-Serrano M. Persistent organic and inorganic pollutants in the effluents from the textile dyeing industries: Ecotoxicology appraisal via a battery of biotests. Environmental Research. 2021;**196**: 110956

[46] Ayalew AA, Aragaw TA. Utilization of treated coffee husk as low-cost bio-sorbent for adsorption of methylene blue. Adsorption Science and Technology. 2020;**38**:205-222

[47] Tran TH, Le AH, Pham TH, Nguyen DT, Chang SW, Chung WJ, et al. Adsorption isotherms and kinetic modeling of methylene blue dye onto a carbonaceous hydrochar adsorbent derived from coffee husk waste. Science of the Total Environment. 2020;**725**: 138325

[48] Shen K, Gondal MA. Removal of hazardous Rhodamine dye from water by adsorption onto exhausted coffee ground. Journal of Saudi Chemical Society. 2017;**21**:S120-S127

[49] Ovando-Medina VM, Dávila-Guzmán NE, Pérez-Aguilar NV, Martínez-Gutiérrez H, Antonio-Carmona ID, Martínez-Amador SY, et al. A semi-conducting polypyrrole/coffee grounds waste composite for rhodamine B dye adsorption. Iranian Polymer Journal. 2018;**27**:171-181

[50] Atirah Mat SS, Syed Zuber SZH, Enche Ab Rahim SK, Ahmad Sohaimi KS, Abdul Halim NA, Zainudin NF, et al. Malachite Green adsorption by spent coffee grounds. IOP Conference Series: Materials Science and Engineering. 2018;**318**:012015

[51] Murthy KTP, Gowrishankar BS, Chandra Prabha MN, Kruthi M, Hari KR. Studies on batch adsorptive removal of malachite green from synthetic wastewater using acid treated coffee husk: Equilibrium, kinetics and thermodynamic studies. Microchemical Journal. 2019;**146**:192-201

[52] Taufik R, Mohamad M, Wannahari R, Shoparwe NF, Osman WHW, Teo PT, et al. Spent coffee ground as low-cost adsorbent for congo red dye removal from aqueous solution. IOP Conference Series: Earth and Environmental Science. 2021;**765**: 012089

[53] Cheruiyot GK, Wanyonyi WC, Kiplimo JJ, Maina EN. Adsorption of toxic crystal violet dye using coffee husks: Equilibrium, kinetics and thermodynamics study. Scientific African. 2019;**5**:e00116

[54] Loulidi I, Boukhlifi F, Ouchabi M, Amar A, Jabri M, Kali A, et al. Kinetic, isotherm and mechanism investigations of the removal of Basic Violet 3 from water by raw spent coffee grounds. Physical Chemistry Research. 2020;**8**:569-584

[55] Pagalan E Jr, Sebron M, Gomez S, Salva SJ, Ampusta R, Macarayo AJ, et al. Activated carbon from spent coffee grounds as an adsorbent for treatment of water contaminated by aniline yellow dye. Industrial Crops and Products. 2020;**145**:111953

[56] Laksaci H, Khelifi A, Belhamdi B, Trari M. The use of prepared activated

carbon as adsorbent for the removal of orange G from aqueous solution. Microchemical Journal. 2019;**145**: 908-913

[57] Jahin HS. Adsorption of Methyl Red from aqueous solutions using activated carbon prepared from coffee residue. International Journal of Environment. 2014;**3**:126-132

[58] Zuorro A, Di BA, Lavecchia R. Magnetically modified coffee silverskin for the removal of xenobiotics from wastewater. Chemical Engineering Transactions. 2013;**35**:1375-1380

[59] Nitayaphat W, Jintakosol T, Engkaseth K, Wanrakakit Y. Removal of methylene blue from aqueous solution by coffee residues. Chiang Mai Journal of Science. 2015;**42**:407-416

[60] Orfanos A, Manariotis ID, Karapanagioti HK. Sorption of Methylene blue onto food industry byproducts. International Biochar Initiative. 2015. Article ID 7046

[61] Pavlović MD, Nikolić IR, Milutinović MD, Dimitrijević-Branković SI, Šiler-Marinković SS, Antonović DG. Plant waste materials from restaurants as the adsorbents for dyes. Hemijska Industrija. 2015;**69**:667-677

[62] Fasfous II, Farha NA. Removal of Cibacron Brilliant Yellow 3G-P Dye from aqueous solutions using coffee husks as non-conventional low-cost sorbent. International Journal of Chemical and Molecular Engineering. 2012;**6**:908-914

[63] Pavlović MD, Nikolić IR, Milutinović MD, Dimitrijević-Branković SI, Šiler-Marinković SS, Antonović DG. Rapid cationic dye adsorption on polyphenol-extracted coffee grounds—A response surface methodology approach. Journal of the Taiwan Institute of Chemical Engineers. 2014;**45**:1691-1699

[64] Kyzas GZ, Lazaridis NK, Mitropoulos AC. Optimization of batch conditions and application to fixed-bed columns for a sequential technique of total color removal using "greek coffee" residues as materials for real dyeing effluents. Journal of Engineering Science and Technology—Review. 2012;**5**:66-75

[65] Ahalya N, Chandraprabha MN, Kanamadi RD, Ramachandra TV. Adsorption of Fast Green on to Coffee Husk. Journal of Chemical Engineering and Research. 2014;**2**:201-207

[66] Al-Ghouti MA, Da'ana DA. Guidelines for the use and interpretation of adsorption isotherm models: A review. Journal of Hazardous Materials. 2020;**393**:122383

[67] Langmuir I. The constitution and fundamental properties of solids and liquids. part i. solids. Journal of the American Chemical Society. 1916;**38**: 2221-2295

[68] Freundlich H. Over the adsorption in solution. Zeitschrift für Physikalische Chemie. 1907;**57U**:385-470

[69] Temkin IM. Kinetics of ammonia synthesis on promoted iron catalysts. Acta Physiochim. 1940;**12**:327-356

[70] Dubinin MM. The equation of the characteristic curve of activated charcoal. Proceedings of the USSR Academy of Sciences. 1947;**55**:327-329

[71] KrishnaTP M, Gowrishankar BS, Krishna RH, Chandraprabha MN, Mathew BB. Magnetic modification of coffee husk hydrochar for adsorptive removal of methylene blue: Isotherms, kinetics and thermodynamic studies. Environmental Chemistry and Ecotoxicology. 2020;**2**:205-212

[72] Wang J, Guo X. Adsorption kinetic models: Physical meanings, applications, and solving methods. Journal of Hazardous Materials. 2020;**390**:122156

[73] Lagergren S. Zur Theorie der sogenannten Adsorption gelöster Stoffe. Zeitschrift für Chemie und Industrie der Kolloide. 1898;**24**:1-39

[74] Ho YS, McKay G. Pseudo-second order model for sorption processes. Process Biochemistry. 1999;**34**:451-465

[75] Devi GK, Kumar KS, Parthiban R, Kalishwaralal K. An insight study on HPTLC fingerprinting of Mukia maderaspatna: Mechanism of bioactive constituents in metal nanoparticle synthesis and its activity against human pathogens. Microbial Pathogenesis. 2017; **102**:120-132

[76] Weber WJ, Morris JC. Kinetics of adsorption on carbon from solution. Journal of the Sanitary Engineering Division. 1963;**89**:31-59

[77] Thomas B, Shilpa EP, Alexander LK. Role of functional groups and morphology on the pH-dependent adsorption of a cationic dye using banana peel, orange peel, and neem leaf bio-adsorbents. Emergent Materials. 2021;**4**:1479-1487

[78] Al-Zaben MI, Mekhamer WK. Removal of 4-chloro-2-methyl phenoxy acetic acid pesticide using coffee wastes from aqueous solution. Arabian Journal of Chemistry. 2017;**10**:S1523-S1529

[79] Kyzas GZ, Lazaridis NK, Mitropoulos AC. Removal of dyes from aqueous solutions with untreated coffee residues as potential low-cost adsorbents: Equilibrium, reuse and thermodynamic approach. Chemical Engineering Journal. 2012;**189**:148-159

Chapter 2

Cellulose Acetate Membrane Preparation for Wastewater Treatment

Ibtissem Ounifi, Fatma Khaled, Malak Kahloul, Hanen Ajari and Amor Hafiane

Abstract

For a long time, humans have used cellulose, as a natural, renewable, and transformative polymer, for scientific development to create new technologies. Cellulose is the most abundant biopolymer on Earth, accounting for more than 50% of terrestrial biomass. For this reason, the treated cellulose (cellulose acetate (CA)) was used in the membrane preparation for water desalination. However, membrane preparation has recently attracted big attention of several research groups. In this case, cellulose acetate (CA), as an inexpensive hydrophilic biopolymer, was chosen as a polymer for preparing the membranes via the inversion phase, since it offers an efficient purification benefit with low energy consumption and less cost. The purpose of this chapter is to describe the various types of membrane preparation based on cellulose acetate, with pathogens, bacteria, and heavy metal (cadmium), and the applications of these membranes in the treatment of contaminated water, to ensure a clean water supply for both human and industrial uses.

Keywords: cellulose acetate, membrane preparation, water desalination, hydrophilic, wastewater treatment

1. Introduction

In the last decades, the population, industrial, agriculture, and urbanization growth has continually increased and by adding climate change and policymaking problems, the fresh and accessible water has been reduced in the world. As a result of global problems, the demand for pure water increased more and more [1, 2]. As well as the problem of heavy metal contamination of water supplies, which is producing major environmental and health implications, there is a growing need for water desalination technology [3–5].

Recently, membrane separation technology has been attracting considerable attention due to its potential applications over conventional approaches such as thermal desalination. Membrane process becomes widely used in water desalination, wastewater treatment, pharmaceutical filtration, and even in energy generation due to their different properties such as being environmentally friendliness, simple operation, high efficiency, low operating cost, low energy use, and scalability [5–7].

Generally, the basic function of the membrane is to separate different mixtures by concentrating small molecules and allowing others to pass. This specific function as well as the other advantages makes membrane technology the principally used application in many separation membrane processes such as water, wastewater treatment, and adsorption, which gives it an important part in the water challenge (**Figure 1**) [8, 9].

Membrane separation is based on an efficient and scalable methodology by using some or without no harmful chemicals by-product formation. Membrane fabrication from synthetic polymers becomes very much used in various separation membrane applications. Generally, several polymers are used such as polysulfone (PSF), polypropylene (PP), polyethylene (PE), polyethersulfone (PES), polyvinylidene fluoride (PVDF), polyvinyl alcohol (PVA), chitosan (CS), and cellulose acetate (CA) [10–13]. Nowadays, CA polymers attract a big interest in membrane fabrication due to their biodegradable polymer propriety, hydrophilicity nature, and large disposability, since cellulose is an organic compound widely existing on the earth [14, 15]. Generally, cellulose can be found in plants, wood, algae, tunicates, and even in bacteria, and its annual extracted production can attain around 7×10^{10} tons per year, which highly encourages the use of cellulose [16]. As result, many researchers focused on membrane fabrication with the derivatives of cellulose such as cellulose acetate CA [14, 15].

The CA membrane's performance was improved by its high chemical and mechanical stability, high hydrophilicity with desalination pilot, high transport properties, low protein adsorption, excellent water affinity, and excellent film-forming property [17]. Due to their desirable properties and environmental aspects, CA-based membranes have been used in several separation membrane applications, such as gas separation [18], reverse osmosis (RO) [19], nanofiltration (NF) [20], ultrafiltration

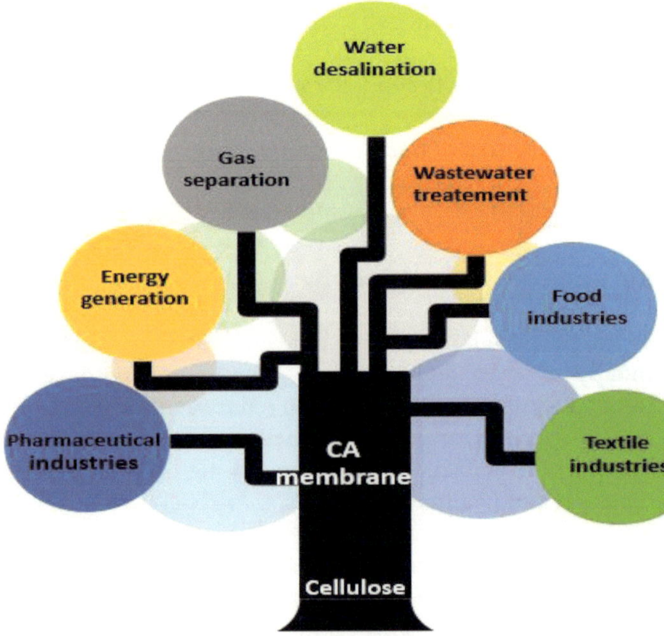

Figure 1.
CA-based membrane applications.

(UF) [21], microfiltration [22] (MF), pervaporation [23], and ion exchange [24], and it has been widely used to electrospun fibers for it simple processing [25].

More than separation membrane technology, CA has the potential to be used securely in industrial and biomedical applications such as wound dressing [26] because of its low toxicity and biodegradability [27].

The wide use of CA membranes can be attributed to the fact of the importance of CA in the manufacturing of biodegradable membranes, besides the possibility of improving membrane performance through the addition of additives [28], particularly nanomaterials. Therefore, to improve the CA membrane's performance, the membrane solution can be mixed with the right organic or inorganic additives in the phase inversion technique [29]. To modify CA membranes, some researchers changed the solvent in the membrane casting solution. Various solvents were tentatively used to prepare CA casting solution, including N-methyl pyrrolidone (NMP), N, N-dimethyl acetamide (DMAc), and a composed solvent of N, N-dimethylformamide (DMF), acetone, or 2-propanol [30–32].

This chapter aims to present the cellulose properties and its derivatives and to study the use of CA polymer in the preparation of polymeric separation membranes applied in different applications. First, a presentation of CA properties will be carried out. Then, different CA membrane configurations will be described. Thus, the performance of the fabricated CA membrane in those applications will be studied and discussed, by taking the CA as the used polymer on the one hand and as an additive on other hand.

2. Cellulose acetate properties

Cellulose acetate is an important cellulose derivative due to its properties, which are used in different applications. It is extracted from natural and renewable resources. Further, CA is a natural polymer that is characterized by flexibility, biodegradability, non-poisonous, and eco-friendly [33]. Due to its important and unique benefits of biodegradability and nontoxic behavior [27], CA has been used frequently in membrane manufacturing, especially in the treatment of water and wastewater technologies. Several polymers have been used in the membrane synthesis, but the CA membranes are the most preferred according to their durability, low cost, and especially required hydrophilic and biodegradability properties [34, 35].

Also, by dint of their chemical and thermal stability (**Figure 2**), good water retention capacity, and great resistance to methanol permeation, the CA membranes are used in many other applications such as gas separation for CO_2 capture or nitrogen generation [36] and fuel separation for fossil fuel and natural gas drainage [37].

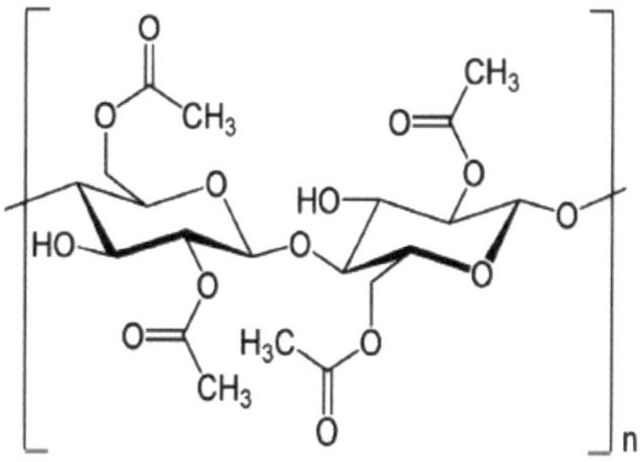

Figure 2.
Chemical structure of cellulose acetate.

Several molecular weights of CA were used in membranes synthesis, which are usually attained at 30,000 g mol^{-1} [38], 37,000 g mol^{-1} [39], 50,000 g mol^{-1} [40], and 60,000 g mol^{-1} [41], between 39 and 40% of this weight is acetyl.

One of the main properties of CA is the possibility of dissolution in green solvents with the ability to use in the phase inversion technique for membrane fabrication [42, 43]. Totally, bio-based NF membranes were fabricated using CA as a base membrane and several glycerol derivatives (monoacetin, diacetin, triacetin, and glycerol-formal) as green solvents 2-methyl tetrahydrofuran as a co-solvent to reach rejection in the NF range [44]. The greatest membranes resulted from diacetin solvent by permeability in the range of 5.5–12.8 L m^{-2} h^{-1} bar^{-1} and >90% Rose Bengal rejection.

3. Configuration of CA membranes

3.1 Flat sheet

Flat sheet membranes in general take the shape of flat sheets. The membrane's top surface is an active layer for fluids or particulate/fluid separation. The membrane surface is the top upper layer that performs the function of separation. However, flat membranes are often prepared by pouring the polymeric solution onto a substrate. The most important characteristics of this type of membrane are high specific flow rates, high production rates, and easy placement of the membrane in the module. There are several approaches for preparing flat sheet membranes, but generally phase inversion or thin film composite and interfacial polymerization are the essential methods used to manufacture these membranes.

3.1.1 Phase inversion

Generally, asymmetric porous polymer membranes are developed by the phase inversion process. This occurs when an initially homogeneous polymer solution changes its conditions thermodynamically.

During phase inversion, a single-phase casting solution causes the formation of two different phases: a polymer-rich phase or a solid membrane matrix and a poor phase in terms of polymer or liquid polymer [45]. This technique enables the change from a liquid to a solid state. The transition from one liquid to two liquids starts the solidification process. Slowly, the rich polymer phase forms the solid matrix during this liquid-liquid transition. The poor phase is then removed by successive washing steps and leaves room for the pores of the membrane. The morphology of a membrane has a significant impact on its characteristics and performance.

Therefore, membrane synthesis's major objective by phase inversion process is to obtain membranes with specific properties such as microfiltration, nanofiltration, or ultrafiltration that ensure good thermal, chemical, and mechanical resistance properties and to develop adequate membranes properties with a surface layer for reverse osmosis and gas separation [46].

The CA membranes prepared by using this technique have been used as microfiltration [46], ultrafiltration [47], nanofiltration [42], forward osmosis [48], reverse osmosis [49, 50], and gas separation membranes [51].

In order to create an asymmetric microfiltration membrane structure, a selective layer and a prefilter layer were combined in one membrane to examine the effect of the CA-based solvent system on membrane morphology during membrane manufacturing. A pseudo-ternary mixture of CA, methyl formate (MF), 2-propanol, and water was used to create asymmetric microfiltration of CA membranes by using a single-layer dry casting process. The created membranes were composed of uniform pore size isotropic regions that are covered by protective skin.

The selective layer of these newly created CA microfiltration membranes fronted the solid-liquid interface in contrast to other asymmetric membranes, or the upper area facing the liquid-gas interface showing larger pore sizes. The denser bottom layer acted as a selective layer while the upper layer functioned as a prefilter. The membranes' structural analyses were carried out using scanning electron microscopy. A theory regarding the actions occurring in the casting solution during membrane development is addressed in depth along with the membrane structures that have been observed [52].

Using the phase inversion method, Ounifi et al. [53] successfully synthesized ultrafiltration membranes from a casting solution containing CA at concentrations ranging from 15 to 21 wt.%. With the rise in CA, the membrane's performance, such as permeability, contact angle, and water content, decreased, but the PEG rejection rose at higher polymer concentrations during membrane formation (**Figure 3**).

CA-NF membranes [54] were prepared via the NIPS inversion process. The asymmetric structure of CA-NF membranes, which are composed of a dense top layer supported by a porous sublayer, was clearly elucidated by the SEM images (**Figure 4**). By increasing the CA concentration in the polymeric doped solution used for membrane synthesis, membrane characterization demonstrated the membrane's hydrophilic nature and decreased porosity. The membrane's permeability to pass water through it was decreased due to reduced porosity, while the salt retention was increased. The size exclusion mechanism dominates the transport mechanism through the membrane.

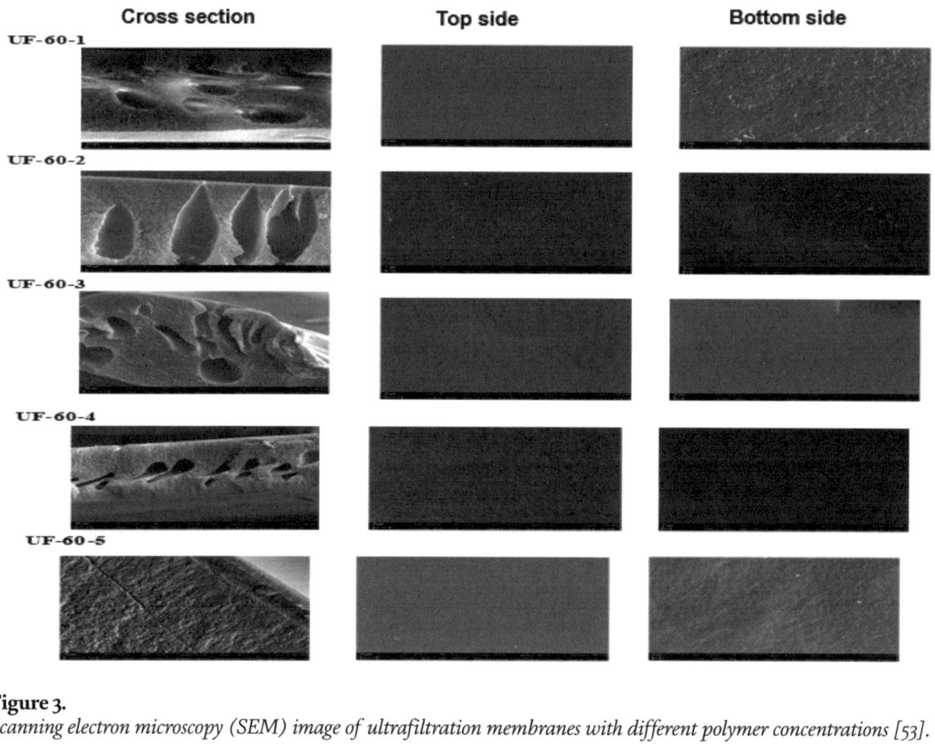

Figure 3.
Scanning electron microscopy (SEM) image of ultrafiltration membranes with different polymer concentrations [53].

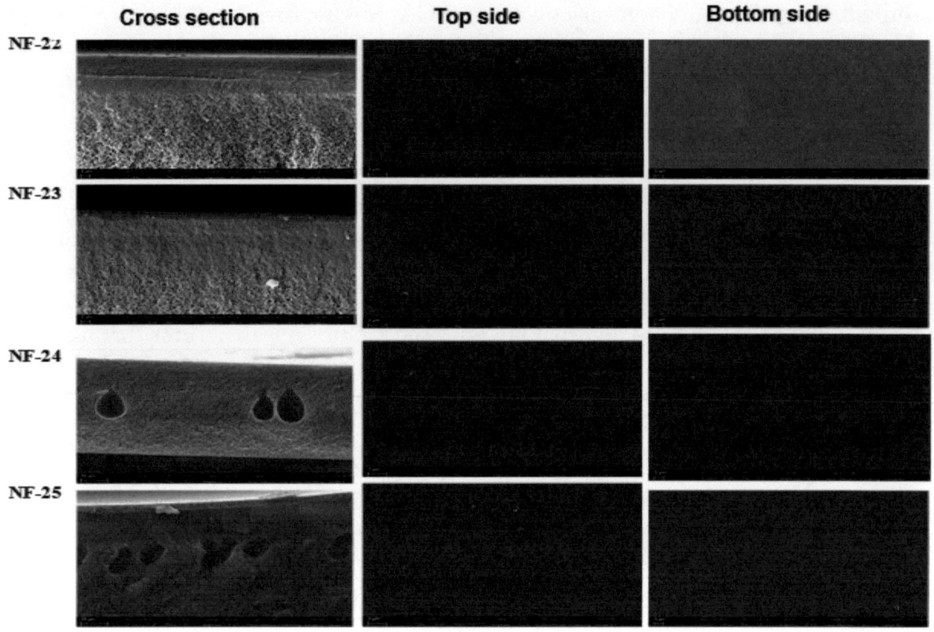

Figure 4.
SEM image of nanofiltration membranes with different polymer concentrations [54].

The order of rejection is determined by the hydration energy and hydrated radii of the ions and is as follows: R(Na_2SO_4) > R($CaCl_2$) > R(NaCl).

In numerous studies, the phase inversion method used in the CA membranes preparation is carried out in conjunction with the inclusion of certain additives such as composites, nano-materials, surfactants, and polymer additives to change the prepared membranes with this approach.

TiO_2 nanoparticles were dispersed in the CA casting solutions to prepare hybridCA/TiO_2 membranes by phase inversion. Investigations were done into how TiO_2 affected the morphology and thermal stability of CA/TiO_2 membranes. According to SEM findings, the addition of TiO_2 nanoparticles makes the hybrid membrane more porous due to increase in the average pore size. The TGA results show that the interaction between TiO_2 nanoparticles and CA exists. Thus, the hybrid membrane's thermal stability was enhanced by the addition of TiO_2 nanoparticles. In addition, PWF membranes show that the addition of TiO_2 nanoparticles leads to an increase in water permeation [49].

CA-silica (SiO_2) hybrid membranes (CA-SiO_2) were synthesized by promoting in situ condensations between silanols of the SiO_2 precursor and COH or acetate groups of the CA polymer. The results showed molecular water strongly linked to hydrogen bonds with SiOH groups, which leads to a drastic drop in the membrane hydraulic permeability, from 57 to 10 kg h^{-1} m^{-2} bar^{-1}. In comparison with the CA membrane, the inclusion of 5 and 10 mol% silica enhanced the hydraulic permeability from 32 to 82 kg h^{-1} m^{-2} bar^{-1} [55].

In order to remove Cd^{2+}, Ounifi et al. developed fouling-resistant UF membranes. Phase inversion was used to prepare the UF membranes by combining PAA and CA. The strong hydrogen interactions between the two polymers were observed with FT-IR. The membranes showed highly porous and asymmetric structures with affected morphologies as a result of PAA concentration. In fact, by increasing the PAA concentration from 0 wt.% to 15 wt.%, the porosity increased from 44.5% to 75.6%. Contact angle and pure water flow measurements showed that adding PAA to the CA structure improved the membrane hydrophilicity. The result ultimately demonstrated that the prepared membrane with a higher PAA content (M15) showed superior permeability and HA rejection characteristics [56].

3.1.2 Interfacial polymerization

Interfacial polymerization (IP) is a very practical and useful simple technique to create a thin layer for nanofiltration and reverse osmosis membranes. Interfacial polymerization always has been of a particular interest in industrial manufacturing.

The thin layer on the membrane surface is formed by the reaction and co-polymerization of two reactive monomers. In recent years, the development of thin-film composite by interfacial polymerization has received attention due to significantly improved membrane properties such as selectivity and fouling resistance. This method involves several steps, and it is easy to apply. To create the hydrophilic NF membranes, the substrate, typically used in MF or UF (cellulose acetate, polysulfone, etc.), must first be dissolved in a solution containing a highly reactive, water-soluble, and bifunctional or trifunctional monomer. In the second step, the membrane is submerged in a solution comprising an additional mono- or polyfunctional monomer that is soluble in an organic water-miscible solvent. The two monomers thus combine to form a dense polymer layer at the water/organic solvent interface. To finish the polymerization reaction and cross-link the resulting structure, this layer is then frequently heated. The layer that forms on the surface of the membrane limits the progression of the reagents, resulting in a dense thin layer of around 50 nm. This is an advantage of interfacial polymerization.

The porous layer is usually made from PSF polymer. However, interfacial polymerization can also be performed on CA-based support, or CA can be used as a selective layer on a support.

According to Ounifi et al., the interfacial polymerization of M-phenylenediamine (MPD) and trimesoyl chloride (TMC) on CA-UF successfully produced innovative flat-sheet thin film composite membranes for nanofiltration (TFC-NF) (**Figure 5**). The thin film composite TFC had a smooth surface, as evidenced by the SEM and atomic force microscopy (AFM) pictures. The polyamide thin top layer preparation conditions, such as reaction time and TMC concentration, have a significant impact on the membrane performance, specifically the water permeability and salt rejection of the TFC-NF. When TMC concentration and IP reaction time were increased, the water permeability fell from 23.35 to 10.44 L m^{-2} h^{-1} bar^{-1}. Additionally, the salt rejections of these TFC membranes were adjusted from 69.23% to 92.43% and from 19% to 55.54% by using Na_2SO_4 and NaCl, respectively. A cadmium solution retention test has been done. The test result showed that this type of membrane could be used for the removal of heavy metals. The rejection of cadmium depends on the nature of the associated anion: $(Cd(NO_3)_2) < R(CdCl_2) < R(CdSO_4)$, and it can reach a maximum of 97.76% for $CdSO_4$ [57].

According to Ounifiet al., the interfacial polymerization technique was used to successfully prepare a number of polyamide TFC membranes. The CA-UF was used as a substrate. The PA layer over the CA-UF substrate was created by using the monomers of 1,3- cyclohexanebis (methylamine) (CHMA) and TMC in the aqueous phase and organic phase, respectively (**Figure 6**). Many concentrations of CHMA, namely 0, 0.2, 0.5, 1, and 2 wt.%, were used on membranes preparation. (The CHMA concentration

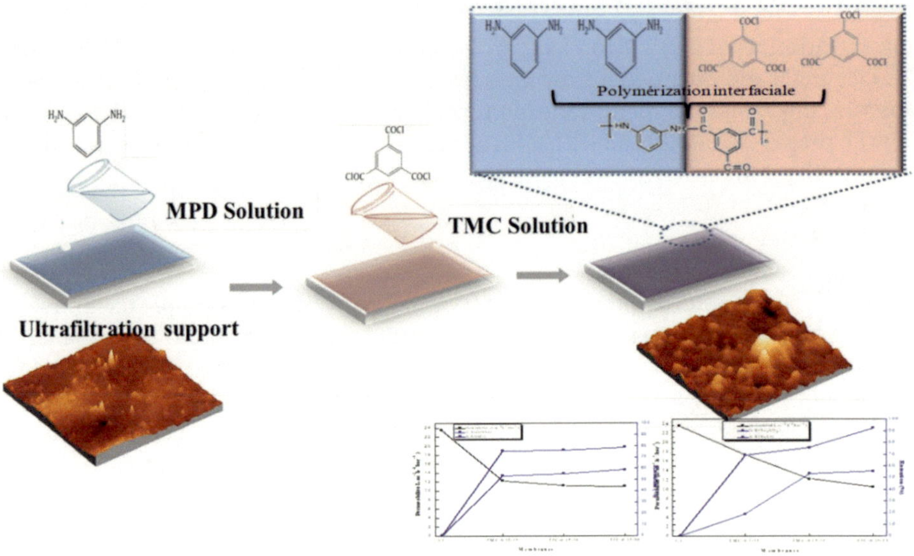

Figure 5.
Mechanism of polymerization interfacial and AFM image of membrane CA and TFC.

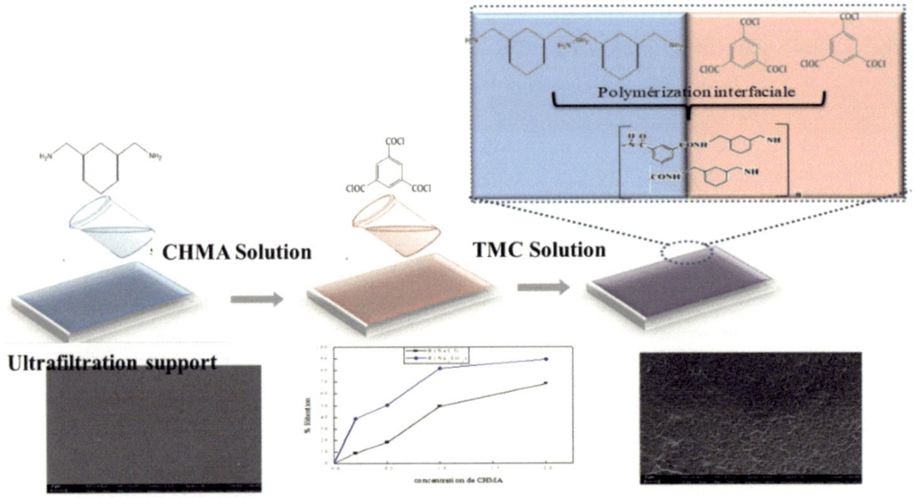

Figure 6.
Mechanism of polymerization interfacial and SEM image of membrane CA and TFC.

effect on the membranes' physicochemical characteristics was investigated by using SEM, FT-IR, water absorption, water permeability, porosity, contact angle, specific salt rejection rate, and dye molecules.) SEM analyses showed that increase of the CHMA concentration increased the membrane's rugosity and thickness. The TFC-2 membrane exhibited the typical polyamide membrane morphology identified by the "ridge and valley" structure. The FT-IR findings revealed novel absorption bands specific to polyamide, which ensure the successful use of interfacial polymerization. The hydrophilicity of the membrane was enhanced by the increase in CHMA concentration, but the porosity and permeability of the membrane decreased. For example, the membrane water permeability varied from 36.02 for the CA-UF membrane to 17.09 L h^{-1} m^{-2} bar^{-1}

for TFC-2. All TFC membranes had a higher rate of salt and dye rejection than the CA-UF membrane. The TFC-2 membrane demonstrated a desalination rate of up to 89%, whereas the malachite green and Congo red were rejected at rates of 89 and 85%, respectively. The concentration of the molecule and the feed solution's pH affect the separation performance. Overall, the CHMA could be a potential candidate to develop TFC membranes for desalination and water treatment [58].

Ounifi et al. studied the influence of support pores size and monomer type on the thin-film composite membranes performance. By using CA membrane as support and the cyclohexane-1,3,5tricarbonylchloride (HTC) monomer as an organic phase, a series of polyamide-cellulose acetate thin-film composite nanofiltration (TFC-NF) membranes were created for this purpose by the interfacial polymerization technique. The effect of the cellulose acetate concentration of 15, 16.5, 18, 19.5, and 21.5% on the pore size of the membrane support was studied. As monomers, m-phenylenediamine (MPD), piperazine (PIP), and 1,3-cyclohexanebis (methylamine) (CHMA) were investigated. In comparison with other produced membranes, the results showed that the PIP/HTC membrane is more hydrophilic and has a more intense granular and convex structure with a rougher surface. The concentration of cellulose acetate affected both the porosity and the water flow. The reduction in porosity was consistent with the results of the SEM analysis (**Figure** 7), contact angle, and permeability. The retention of NaCl, $CaCl_2$, and Na_2SO_4 was evaluated for the membrane performances. Comparing the Na_2SO_4 rejection with $CaCl_2$ and NaCl, the former is higher than the latter [55].

3.2 Hollow fiber

Another membrane configuration, in which CA can be used as the main polymer, is a hollow fiber membrane. The hollow fiber membrane (HF) has three main advantages, which have led to the hollow fiber design becoming a popular choice:

1. Compared with flat and spiral-wound modules, hollow fiber modules offer a substantially higher ratio of membrane area to unit volume (several thousand M2/M3), resulting in higher production per unit volume of membrane module;

2. They may be back-washed to recover the flux due to their self-supporting;

3. They also have good operational flexibility [59].

The two main approaches for preparing HF are thermally induced phase separation (TIPS) and non-solvent-induced phase separation (NIPS). For NIPS, the solvent is extracted from the solution using a non-solvent, which is a substance that is miscible with the solvent but cannot solubilize the polymer. When the solvent is removed from the solution, the polymer concentration rises to its solubility limit, causing precipitation. The membrane structure can be tuned by adjusting preparation parameters such as polymer concentration, non-solvent system strength, and the use of additives [60].

For TIPS, phase separation is achieved by lowering the temperature of the dope solution. To obtain a homogeneous solution, the polymer is dissolved in one or more solvents, commonly referred to as diluents, at a high temperature. The dope solution is then extruded at high temperature through a spinneret and cooled in a low-temperature-quenching bath. At low temperatures, the diluents are unable to solubilize the polymer, and the solution undergoes phase separation [61]. After the hollow fiber membrane is formed, the remaining solvent is removed with cleaning baths [62].

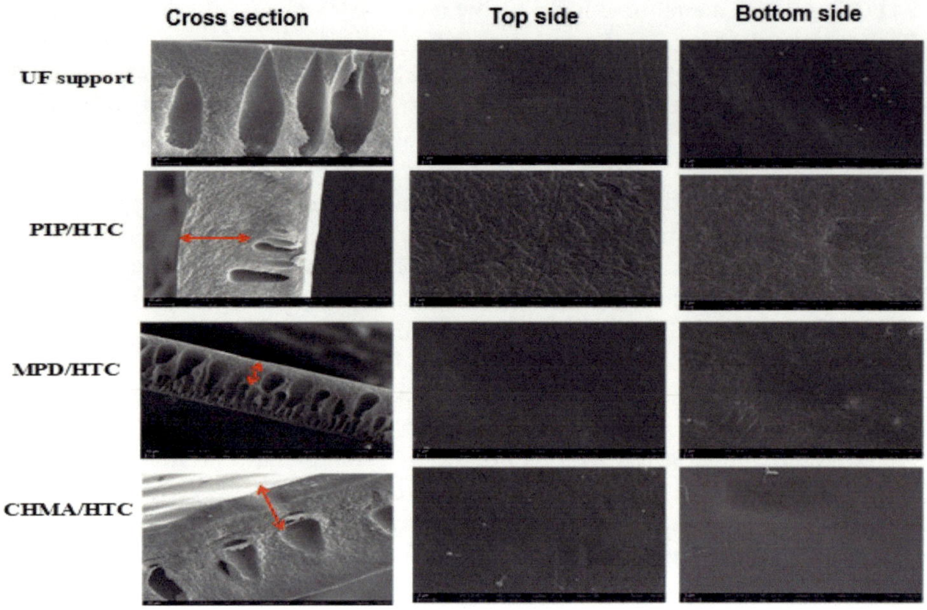

Figure 7.
SEM image of TFC nanofiltration membranes with different monomers [55].

The diluent properties and solution viscosity both play a significant role in determining the final membrane morphology.

A homogeneous braid-reinforced CA hollow fiber membrane was formed through a non-solvent-induced phase separation process, and the membrane was prepared using an HMR method, in which CA solution was coated on the outer surface of the CA braid [63].

CA hollow fiber membranes were fabricated via various spinning processes such as dry-jet wet spinning [64], well-known dry-jet wet spinning process, dry-wet phase inversion method [65], coating a separation layer on the high-strength hollow tubular braid, melt spinning, and stretching process [66].

Most CA hollow fiber membranes were prepared using the dry-wet phase inversion method (**Figure 8**), but their poor mechanical properties limited their use in engineering practice. As a result, a hollow fiber membrane with high mechanical properties was required. Many studies have been conducted in order to improve the mechanical properties of hollow fiber membranes. For example, coating a separation layer on the high-strength hollow tubular braid was a relatively simple and effective method. There were two interfaces between the braid and the separation layer: the homogeneous-reinforced (HMR) interface and the heterogeneous-reinforced (HTR) interface. According to Fan et al. [63], the homogeneous-reinforced (HMR) method and the heterogeneous-reinforced (HTR) technique were combined to reduce the weaknesses of each and achieve the best results. Fan et al. produced a BR CA hollow fiber membrane by coating the "hybrid" braid with CA solution. The two-dimensional braided technique was used to prepare the "hybrid: braid, which consisted of CA filaments as the homogeneous fiber and PAN filaments as the heterogeneous fiber as shown in **Figure 9** [63]. SEM image results illustrated the cross section of BR CA membranes made by CA filaments and PAN filaments process in various braid compositions as shown in **Figure 10**.

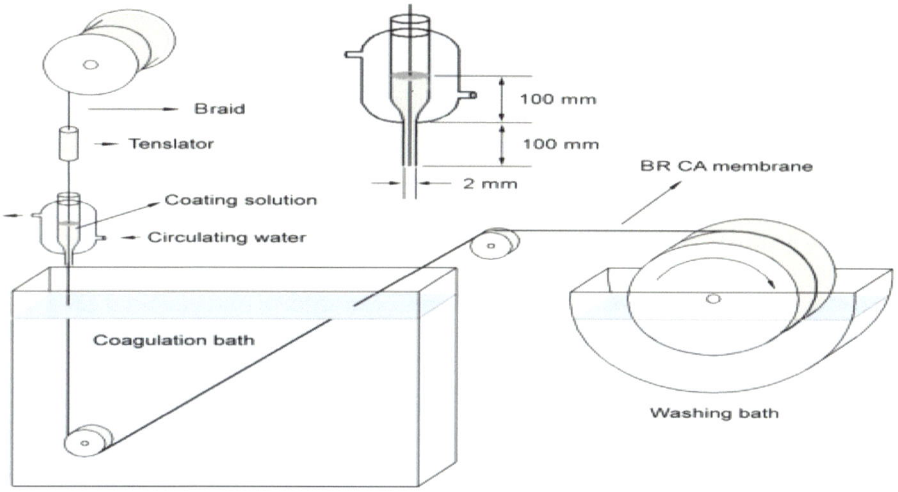

Figure 8.
Spinning apparatus set up [63].

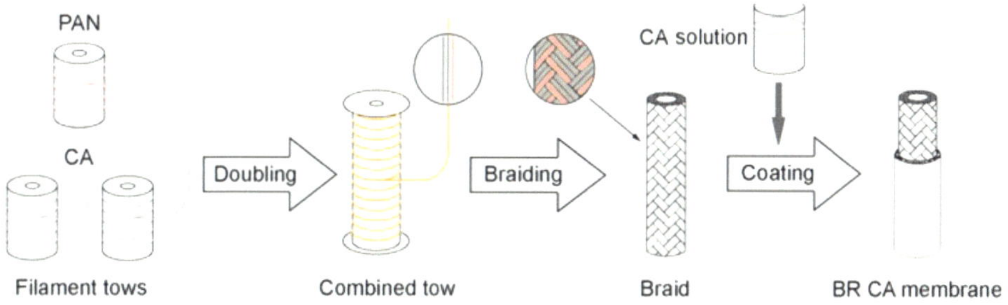

Figure 9.
BR CA membrane process with CA filaments and PAN filaments.

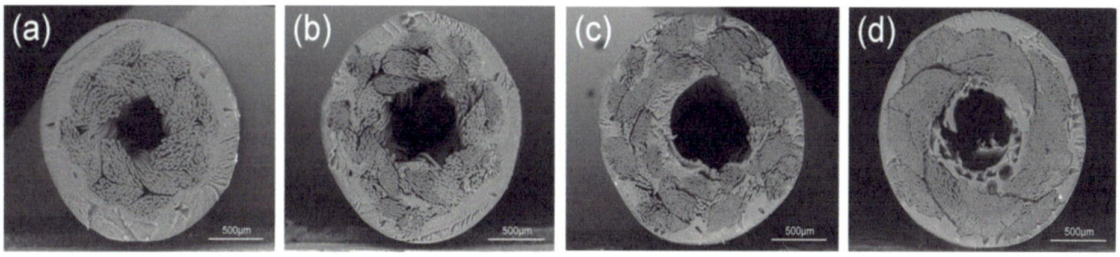

Figure 10.
SEM images of cross section of BR CA membranes with various braid compositions.

The mechanical properties and permeation performance of BR CA hollow fiber membranes with different fiber ratios in the braid and different CA concentrations in the coating solution were investigated, giving a new approach to obtaining BR CA membranes with desirable performance.

The potential and application of commercial hollow fiber membrane products are majorly affected by three factors:

1. The thickness of the functional separation or selective layer and its substructure morphology;

2. The pore size and pore-size distribution of the functional separation or selective layer;

3. The chemistry, mechanical, and physicochemical characteristics of the membrane material.

CA-based NF hollow fiber membranes have been studied and explored for FO applications. Heat treatment at 60°C for 60 min slightly affects the membrane pore radius and the thickness of the outer skin layer. Further, heat treatment at 95°C for 20 min results in the significant shrinkage of pores on the membrane surface with a denser outer skin layer. The resultant fiber has a high rejection to NaCl and $MgCl_2$ and low pure water permeability in the NF tests. The CA-NF hollow fiber heat-treated at 95°C shows promising FO performance. However, this type of CA-NF hollow fiber membrane shows a decrease in performance ratio with increasing saline concentration in the feed due to the more severe internal concentrative concentration polarization. These preliminary results manifest the potential of the CA NF hollow fiber membranes for concentrating various feed solutions in the FO process. Future works should aim at the reduction of the dense layer thickness to enhance the flux while maintaining a low salt leakage rate [67].

3.3 Electrospinning

Electrospinning has drawn much attention and has already been as one of the most important techniques for manufacturing nonwoven polymer nanofiber membranes (synthetic or bio-based) with diameters ranging from nanometers to microns. Many geometries of the membrane are available for a diverse range of applications, along with agri-food, biomedical, cosmetics energy, and even textiles. For producing fibers from a polymer solution and synthesizing sophisticated three-dimensional structures, the electrospinning process employs thousands to tens of thousands of volts of high-voltage electrostatic repulsion [68]. A large number of synthetic polymers, including polycaprolactone (PCL), poly-lactic acid, and poly(lactic-co-glycolic acid), as well as more than 200 different types of natural polymers and composites, such as gelatin, silk fibroin, chitosan, and collagen, have been used in electrospinning technology systems. With this technology, the synthesized nanofibers have a high surface area-to-volume ratio, excellent plasticity, and flexibility, which can improve cell adhesion, proliferation, and differentiation activities. Lyu and his coworkers [69] studied the electrospun cellulose acetate (CA) membranes with optimized porous structures. They demonstrated that the synergy between pore size and porosity is extremely crucial for moisture-induced electricity generation from porous AC membranes. They will also prove that the pore size and porosity of CA membranes can be easily adjusted using a compression process. Several investigations are made to apply the membranes manufactured by electrospinning. For example, Cheng et al. [70] developed a new type of deacetylated cellulose acetate (DA)-polydopamine (PDA) composite nanofiber membrane by electrospinning. The membrane was employed as a highly efficient adsorbent for removing methylene blue (MB) from an aqueous solution. The results demonstrated the successful development of a uniform PDA coating layer on the surface of DA nanofibers. After 30 h of adsorption, the

DA-PDA nanofiber membrane had an adsorption capacity of 88.2 mg/g at a temperature of 25°C and a pH of 6.5, which is approximately 8.6 times greater than that of DA nanofibers. The experimental results revealed that the DA-PDA composite nanofiber's adsorption behavior followed Weber's intraparticle diffusion model, the pseudo-second-order model, and the Langmuir isothermal model. According to a thermodynamic analysis, endothermic, spontaneous, and physisorption processes were observed. Furthermore, Acid black 172 was removed from simulated wastewater using electrospun CA/P(DMDAAC-AM) composite nanofibrous membranes [71, 72]. The equilibrium adsorption capacities were 116, 159, and 192 mg/g, respectively, when the proportion of P(DMDAAC-AM) to CA was 20, 30, and 40wt.%. When the proportion of P(DMDAAC-AM) was 40%, the average fiber diameter was 185 nm, and the maximum adsorption capacity could reach 231 mg/g. CA/P(DMDAAC-AM) nanofibrous membrane adsorption capacities had a well-linear relationship with P(DMDAAC-AM) content and fiber diameter.

4. Additives

CA has been widely used in the manufacture of membranes due to its hydrophilicity, adsorption properties for dyes, heavy metals, and low cost. But in order to improve permeability, separation efficiency, and antifouling properties, in many studies, the CA membranes were prepared by adding certain additives such as composites, nanomaterials, surfactants, and additive polymers to modify the membranes [73, 74].

Ounifi et al. [56] developed new UF membranes incorporating polyacrylic acid (PAA) as an additive. The strong hydrogen interactions between the two polymers observed with FT-IR and the presence of a single peak in the DSC results confirmed the homogeneity of the mixture of CA and PAA in the prepared membranes. The membranes presented asymmetric and highly porous structures with morphologies affected by the PAA concentration. Indeed, the porosity and permeability increase proportionally with the concentration of PAA and inversely proportionally with the contact angle. The benefits of blending CA with PAA have also been observed in terms of antifouling properties, allowing efficient flux recovery from the membrane fouled by the treatment.

CA/TiO_2 hybrid membranes were prepared by phase inversion method, by dispersing titanium dioxide (TiO_2) nanoparticles in the CA casting solutions [49]. The influence of TiO_2 on the morphology and thermal stability of CA/TiO_2 membranes has been studied. The SEM results show that the addition of TiO_2 nanoparticles causes the hybrid membrane to become more porous due to the increase in average pore size. The thermogravimetric analysis results show that the interaction exists between TiO_2 nanoparticles and AC. Moreover, the thermal stability of the hybrid membrane was improved by the addition of TiO_2 nanoparticles. PWF membranes show that the addition of TiO_2 nanoparticles leads to an increase in water permeation.

Ultrafiltration membranes based on CA modified with zinc oxide (ZnO) were prepared [75]. Membrane performance was improved. For example, mechanical stability, morphology, contact angle, and porosity were evaluated on modified CA and ZnO samples. The increase in ZnO concentration in the membrane leads to a decrease in permeability and an increase in hydrophobicity compared with the pure CA membrane.

To improve antifouling properties and shedding, PVP has been added to CA to produce a membrane [76]. PVP is a hydrophilic additive polymer and blowing agent.

From the results, it was significant that the membrane mixed with the hydrophilic additive PVP was improved and showed better properties in terms of surface roughness, hydrophilicity, and thermal and mechanical stability.

CA membranes modified with various surfactant additives, including Span-20 (sorbitan monolaurate), Tween-40 (polyoxyethylene sorbitan monopalmitate), and Tween-60 (polyoxyethylene sorbitan monostearate), were prepared [77]. The effect of these additives has been studied. The membranes' behavior depends on different concentrations of surfactant. It was found that even adding small amounts of the studied surfactant additives could affect the separation behavior and cause the formation of macro voids and finger-like structures. In addition, the effect of surfactant hydrophobicity on the contact angle and water permeation flux of the prepared

Polymer	Additives	Preparation technique	Membrane	Application	Ref
CA	Ag-Nps	NIPS	UF	Water flux 1651 L m^{-2} h^{-1}	[78]
CA	SiO_2	NIPS	UF	Permeability increased from 32 to 82 kg h^{-1} m^2 bar^{-1}	[79]
CA	SiO_2	TIPS	Gaz separation	CO2/CH4 selectivity 80	[80]
CA	Fe_3O_4	Vacuum filtration of nanoparticles on commercially CA membrane	NF	Cu^{2+} Rej.: 63.2% Cd^{2+}Rej.: 64.1% Cr^{6+}Rej.: 70.2%	[81]
CA	TiO_2	NIPS	UF	Removal of Chrome (VI)	[82]
CA	ZnO	Casting	UF	Antibacterial activity against *S. aureus*	[83]
CA	ZrO_2	NIPS	Hollow fiber	Arsenic removal of 87.24% and permeability of 89.94 L m^{-2} h^{-1} bar^{-1}	[84]
CA	CNT	NIPS	RO	5 g L1 NaCl Rej.: 96% Flux: 100 L m^{-2} h^{-1} (35 bar, 0.01 wt% CNT	[85]
CA	GO	NIPS	RO	The synthetic seawater salt rejection of the CA membrane modified with 1 wt.% GO at 25 bar was 1.8 times greater than that of the pure CA membrane	[86]
CA	PVA	NIPS	Hemodialysis	42.484 L/m^2 h is the maximum flow of water CA-PVA rejected up to 95% of bovine serum albumin	[87]
CA	PEG	NIPS	UF	The pure water flux elevated from 49.5 Lm^{-2} h^{-1} to 62.2 Lm^{-2} h^{-1} Rejection result for chromium metal of 31.89%	[88]

Table 1.
Some of the additives used with CA in membrane preparation.

membranes was investigated. It was found that the membrane derived from tween-60 has the lowest flux and the highest contact angle (77.2°) (**Table 1**).

5. Application of CA membranes

5.1 Mirofiltration

In a combination with solvents, CA and polyethyleneimine (PEI), which were cross-linked by polyisocyanate, were combined to prepare a modified MF membrane via phase inversion approach. PEI may provide linkage sites for affinity-separated ligands in the membranes or serve as ligands for metal chelation, endotoxin elimination, or ion exchange [46]. The impacts of phase inversion time, induced by water vapor mixed with PEI and cross-linking agent concentrations, were examined in the function of membrane performance. The prepared blend membranes with 0.15 50 wt.% PEI/CA ratio and 0.5 wt.% cross-linking agent/PEI were used to adsorb Cu^{2+} and bovine serum albumin (BSA) separately. The produced membrane has a maximum Cu^{2+} ion adsorption capacity of 7.42 mg/g dry membrane. The maximum BSA adsorption capacities are of 86.6 and 43.8 mg/g dry membranes for the membranes with and without the chelating Cu^{2+} ion, respectively.

Commercial CA-MF membranes were modified to exhibit antibacterial capabilities by being impregnated with thymol mixed in very critical solvents. Batch impregnation was used, and the thymol loading was watched as a result of treatment pressure and time. *Staphylococcus aureus* and *Pseudomonas aeruginosa* were used to test the obtained materials' antibiofilm capacities, while *S. aureus* and *Escherichia coli* were used to test the blockage of membranes in contact with bacteria. The finding's results showed a quick impregnation process that could achieve high thymol loads in just 0.5 h at 15 MPa and 20 MPa. A concentration of 20%Thymol exhibited high antibiofilm effects against the tested strains without changing the membrane performance. The research revealed that the specified polymer could benefit from these potent antibacterial characteristics. These great antibacterial characteristics could be sustainably added to the commercial polymer structure of membranes. The results revealed also that the modification of the commercial cellulose acetate microfiltration membranes with polymeric structure might be successfully accomplished in a quick (30 min) and safe environmental procedure. Therefore, modified membranes with great antibacterial properties may find use in the biomedical sector and venting.

5.2 Ultrafiltration

Ultrafiltration (UF) membrane was prepared by Ounifi et al. [53] by using the Manjikian process by phase inversion with progressively rising polymer concentrations (cellulose acetate at 15, 16.5, 18, and 21 wt.%). At increased polymer concentrations, the membranes showed improved ability to retain pathogenic and epidemiological Uro Pathogenic Escherichia coli (UPEC). The membranes with the highest polymer concentration were able to hold back an entire discharge of 8 U-log bacteria.

By using a blending method, new fouling-resistant UF membranes were prepared for cadmium (Cd) remediation by fusing the adaptability of CA with the intricate characteristics of poly(acrylic acid) (PAA). Since CA/PAA membranes performed better than pure CA membranes in UF experiments, it was clear that the mixture had many advantages. Overall, PAA is a useful additive for creating CA membranes with enhanced antifouling exploitable qualities for the UF treatment of heavy metal-polluted water bodies.

The synthesis, evaluation, and use of polyurethane (PU) with cellulose acetate blend membranes are all part of this research [89]. Numerous characterization methods, including contact angle measurement, ATR-FTIR, water content, and SEM, have been used to investigate the characteristics of PU-CA blend membranes. It was determined that common functional groups were used as raw materials (polyurethane and cellulose acetate) in the PU-CA mixed membrane. It was demonstrated by contact angle measurements and water content values that cellulose acetate's addition decreased the hydrophilicity of the modified membranes, due to the fact that pure polyurethane membrane has higher hydrophilicity than pure cellulose acetate. Aqueous potassium dichromate solution used for ultrafiltration as one of the promising separation techniques was used to test the suitability of the modified membranes. The PU-CA1 membrane was experienced varied chromium ion concentrations for many pH and pressure levels. At pH 3 and 0.4 MPa, process economy was attained for all chromium ion concentrations. The PU-CA1 membrane offers a novel option for the treatment of effluents in the textile sector due to its specific properties.

5.3 Nanofiltration

CA-NF membranes were prepared by using the phase inversion method [54]. The concentration of cellulose acetate was varied from 22 to 25 wt.% to investigate its effect on the characteristics of the membrane. The membrane's characterization revealed that the increase of CA concentration makes it more hydrophilic and decreases the membrane's porosity, consequently reduction in water permeability across the membrane, while it increases salt retention. Knowing that the size exclusion mechanism predominates the transport mechanism through the membrane, the ions rejection was sequenced in the following order: $R(Na_2SO_4) > R(CaCl_2) > R(NaCl)$ as determined by the hydration energy and hydrated radii of the ions. CA NF membranes have been successfully used for cadmium elimination. The rejection rate of removed cadmium reached up to 98%. In terms of steric hindrance and hydration energy, the results revealed that the Cd^{2+} coupled with divalent anions is more rejected than that coupled with monovalent anions. Additionally, the retention reduces from 10^{-4} to 10^{-2} mol L^{-1} when the salt concentration rises. Thus, the obtained results showed the feasibility of using developed CA-NF membranes in removing cadmium from contaminated water.

In order to increase the rejection of Pb^{2+}ions and the permeation flux, vinyl triethoxysilane-graphene oxide/gum arabic was added as filler to modify the CA-based membranes [39]. By using NF membranes, the rejection of Pb^{2+}ions was studied by investigating the effect of contact time, pH, permeation flux, and pressure. It was demonstrated that at pH 9, the rejection reached its optimal with 97.6%, whereas a low pH of 1–6 was not required for salt rejection. The permeation flux was 8.6 l m^{-2} h^{-1}, which is caused by concentration and pressure variations. The special characteristics of the nanomaterial-based membrane and their connection to the current treatment methods offer great prospects to revolutionize water and wastewater treatment as well as to further our knowledge of membrane performance. NF-CA may present a significant factor in the commercial manufacturing of membranes.

5.4 Reverse osmosis

Diainabo et al. aimed to evaluate the pertinence of CA, PSF, and silica (SiO_2) for the treatment of wastewater. For this purpose, PSF-blended CA hybrid filtration membranes (CA/PSF) also PSF-blended and SiO_2-embedded CA hybrid adsorption membranes (CA/PSF-SiO_2) were prepared and characterized for Cu^{2+}, Fe^{2+}, and Zn^{2+} ions elimination from contaminated aqueous solutions. Phase inversion method was used to prepare the membranes by using CA, PSF, and SiO_2 granules that were dissolved in N, N dimethyl formamide (DMF). Different pore sizes were detected at the rough surfaces and cross sections of the membranes, which were studied by using SEM. The obtained membranes' porosity and pore sizes ranged from 26.8 ± 0.3 to 81.1 ± 0.3 μm and from 1.26 to 1.38%, respectively, based on the differentiation. The membranes' contact angles varied from 49 to 76 degrees on their glass side and from 56 to 77 degrees on their air side [90].

More than 90% of the initial metal ions present in the polluted solution, which was concentrated at 40 mg L^{-1}, were taken up by the hybrid filtration polymer membranes. CA/PSF-SiO_2 membranes were used in adsorption tests. These substances were found to have a higher capacity for adsorption than many other well-known adsorbents in the literature, which reach 70 mg g^{-1} for CA/PSF 85/15-SiO_2 for Cu^{2+}.

Membrane type	Membrane properties	Application fields	Retention	Ref
Electrospun CA	Hydroxyapatite (CA/HAp) nanocomposite	Fe^{3+} and Pb^{2+} ions rejection	99.7–95.46%	[91]
Electrospunnanofibrous membranes of CA	Grapheme oxide/magnetite/hydroxyapatite/cellulose acetate nanofibrous membranes	Cr^{6+}, Se^{4+} and methylene blue removal	93.4–95.1%	[92]
CA	Deacetylated cellulose acetate (DA)@polydopamine (PDA)	Methylene blue removal	88.15 mg/g	[93]
CA	CA/MOFDPC (metal-organic frame work derived porous carbon) membranes	Methylene blue removal	87.9%	[94]
CA polymeric membrane	Polymeric membrane impregnated with ZnO nanostructures	Methylene blue removal	75%	[95]
CA	Cellulose acetate/grapheme oxide/sodium dodecyl sulphate (CA/GO/SDS)	phosphate removal	87.2%	[96]
CA	Cellulose acetate/block copolymer membranes	Polyphenols	67.4314 mg/g	[97]
CA based mixed matrix membrane (MMM)	Fe-Al-Mn@chitosan/cellulose acetate	Fluoride de contamination	Excellent defluoridation performance	[98]
CA phthalate (CAP)	Poly(methacrylic acid) (PMAA)/CA	Heavy metals Cu^{2+}, Hg^{2+} and Cd^{2+}	4.8 mg/g	[99]

Table 2.
Some of the application of CA membrane.

Finally, it was discovered that the removal of Cu^{2+}, Fe^{2+}, and Zn^{2+} could be improved by coupling adsorption with ultrafiltration in the tangential mode. This allowed for a removal efficiency of 95% for Cu^{2+} at a low metal concentration of 60 mg L^{-1}, and at a high metal concentration of 900 mg L^{-1}, it showed a great removal efficiency of around 98% [90].

In order to remove metal ions from AMD, the performance of a combined filtration and adsorption process was studied. The prepared membranes showed the ability to extract metal ions from polluted solutions at low concentrations of 40 and 60 mg L^{-1}, whereas at a high concentration of 900 mg L^{-1}, the filtration was ineffective. The selected adsorbents demonstrated a better adsorption capacity than those from other adsorbents used in the literature. The maximum adsorption efficiency with the CA/PSF 85/15-SiO_2 membrane reached between 35 and 70 mg g^{-1} of Cu^{2+}, Fe^{2+}, and Zn^{2+} ions, while the adsorbent had the lowest capacity adsorption with CA/PSA-SiO_2 membrane reached from 20 to 61 mg g^{-1} toward the same metal ions. By combining the effects of the filtration and adsorption membranes, a great metal uptake was obtained at both low and high concentrations. Consequently, this strategy presents a reliable method to remove the metal from aqueous solutions, and it has potential use in industrial wastewater treatment (**Table 2**) [100].

6. Conclusion

Cellulose acetate (CA) as one of the most important derivatives of cellulose was studied. Cellulose acetate has attracted big attention as a polymer due to its desirable properties, which are biodegradability, hydrophilicity, environment friendliness, sustainability, and low cost. In this chapter, the use of cellulose acetate CA in the fabrication of polymeric membranes was studied. CA-based membranes are applied in different fields such as gas separation, pharmaceutical industries, adsorption, water treatment, and wastewater treatment by using separation membrane processes such as nanofiltration, ultrafiltration, microfiltration, and reverse osmosis; in addition, the different configurations of fabricated CA membranes such as flat sheet, hollow fiber, and electrospinning were discussed. Also, the different preparation methods such as phase inverse and interfacial polymerization approaches were outlined. The CA-based membrane has the property to be blending to organic and inorganic additives. To improve the performance of the CA-based membranes and their structural properties, different cost-effective additives were added to the polymeric solution, especially nanomaterials. The obtained CA-based membranes with additives were characterized by their high biodegradability, high hydrophilicity, and high retention of dye molecules and heavy metals.

Abbreviations

CA	cellulose acetate
PS	polysulfone
PP	polypropylene
PE	polyethylene
PES	polyethersulfone
PVDF	polyvinylidene fluoride
PVA	polyvinyl alcohol

CS	chitosan
PCL	polycaprolactone
PDA	polydopamine
MF	microfiltration
IP	interfacial polymerization
NIPS	and non-solvent-induced phase separation
TIPS	thermally induced phase separation
UF	ultrafiltration
NF	nanofiltration
OR	reverse osmosis
HF	hollow fiber membrane
TFC	Thin film composite
NMP	N-methyl pyrrolidone
DMAc	N, N-dimethyl acetamide
DMF	N, N-dimethylformamide
PIP	Piperazine
CHMA	1,3- cyclohexanebis (methylamine)
TFC	M-phenylenediamine
MPD	trimesoyl chloride
HTC	cyclohexane-1,3,5tricarbonylchloride
PEI	polyethyleneimine
PAA	poly(acrylic acid
MB	methylene blue
PU	polyurethane
SEM	scanning electron microscopy
AFM	atomic force microscopy
FTIR	Fourier-transform infrared spectroscopy
TAG	thermogravimetric analysis
HMR	homogeneous-reinforced

Author details

Ibtissem Ounifi[1]*, Fatma Khaled[2], Malak Kahloul[1], Hanen Ajari[2] and Amor Hafiane[1]

1 Laboratory, Water, Membrane and Environmental Biotechnology, Centre of Research and Water Technologies, Technopark of Borj-Cedria, Soliman, Tunisia

2 Laboratory, Energy, Water, Environment and Processes, National School of Engineering of Gabes University of Gabes, Gabes, Tunisia

*Address all correspondence to: ibtissemounifi@yahoo.com

References

[1] Arefi-Oskoui S, Khataee A, Safarpour M, Orooji Y, Vatanpour V. A review on the applications of ultrasonic technology in membrane bioreactors. Ultrasonics Sonochemistry. 2019;**58**:104633. DOI: 10.1016/j.ultsonch.2019.104633

[2] Kobya M, Soltani RDC, Omwene PI, Khataee A. A review on decontamination of arsenic-contained water by electrocoagulation: Reactor configurations and operating cost along with removal mechanisms. Environmental Technology and Innovation. 2020;**17**:100519. DOI: 10.1016/j.eti.2019.100519

[3] Figoli AC, Priyanka M, Sergio S, Jochen B, Jan H, Alessandra C. Innovative Membrane Applications for Arsenic Removal: Innovative Materials and Methods for Water Treatment Solutions for Arsenic and Chromium Removal. London: CRC Press; 2016. p. 22

[4] Song J, Li XM, Ursino C, Figoli A, Hoinkis HTJ. Removal of arsenic by nanofiltration: A case study on novel membrane materials. In: Membrane Technologies for Water Treatment: Removal of Toxic Trace Elements with Emphasis on Arsenic, Fluoride and Uranium. London: CRC Press, Taylor and Francis Group; 2016. pp. 201-214

[5] Safikhani A, Vatanpour V, Habibzadeh S, Saeb MR. Application of graphitic carbon nitrides in developing polymeric membranes: A review. Chemical Engineering Research and Design. 2021;**173**:234-252. DOI: 10.1016/j.cherd.2021.07.020

[6] Santoro S, Galiano F, Jansen JC, Figoli A. Strategy for scale-up of SBS pervaporation membranes for ethanol recovery from diluted aqueous solutions. Separation and Purification Technology. 2017;**176**:252-261. DOI: 10.1016/j.seppur.2016.12.018

[7] Ursino CS, Simone L, Donato S, Santoro MP, De Santo E, Drioli DNE, et al. ECTFE membranes produced by non-toxic diluents for organic solvent filtration separation. RSC Advances. 2016;**6**(84):81001-81012. DOI: 10.1039/c6ra13343f

[8] Warsinger DM, Chakraborty S, Tow EW, Plumlee MH, Bellona C, Loutatidou S, et al. Review of polymeric membranes and processes for potable water reuse. Progress in Polymer Science. 2018;**81**:209-237. DOI: 10.1016/j.progpolymsci.2018.01.004

[9] Orooji Y, Movahedi A, Liu Z, Asadnia M, Ghasali E, Ganjkhanlou Y, et al. Luminescent film: Biofouling investigation of tetraphenylethylene blended polyethersulfone ultrafiltration membrane. Chemosphere. 2021;**267**:128871. DOI: 10.1016/j.chemosphere.2020.128871

[10] Arefi-Oskoui S, Khataee A, JabbarvandBehrouz S, Vatanpour V, HaddadiGharamaleki S, Orooji Y, et al. Development of MoS2/O-MWCNTs/PES blended membrane for efficient removal of dyes, antibiotic, and protein. Separation and Purification Technology. 2022;**280**:119822. DOI: 10.1016/j.seppur.2021.119822

[11] Farnam M, Bin Mukhtar H, Bin MohdShariff A. A review on glassy and rubbery polymeric membranes for natural gas purification. ChemBioEng Reviews. 2021;**8**:90-109. DOI: 10.1002/cben.202100002

[12] Jain H, Garg MC. Fabrication of polymeric nanocomposite forward osmosis membranes for water

desalination. A review. Environmental Technology and Innovation. 2021;**23**:101561. DOI: 10

[13] Vatanpour V, YavuzturkGul B, Zeytuncu B, Korkut S, Ilyasŏglu G, Turken T, et al. Polysaccharides in fabrication of membranes: A review. Carbohydrate Polymers. 2022;**281**:119041. DOI: 10.1016/j.carbpol.2021;119041

[14] Miao X, Lin J, Bian F. Utilization of discarded crop straw toproduce cellulose nanofibrils and their assemblies. Journal of Bioresources and Bioproducts. 2020;**5**(1):26-36. DOI: 10.1016/j.jobab.2020.03.003

[15] Wei David W, Wei H, Gauthier AC, Song J, Jin Y, Xiao H. Super hydrophobic modification of cellulose and cotton textiles: Methodologies and applications. Journal of Bioresources and Bioproducts. 2020;**5**:1-15

[16] Douglass EF, Avci H, Boy R, Rojas OJ, Kotek R. A review of cellulose and cellulose blends for preparation of bio-derived and conventional membranes, nanostructured thin films, and composites. Polymer Reviews. 2018;**58**:102-163. DOI: 10.1080/15583724.2016.1269124

[17] Fischer S, Thümmler K, Volkert B, Hettrich K, Schmidt I, Fische K. Properties and applications of cellulose acetate. Macromolecular Symposia. 2008;**262**:89-96. DOI: 10.1002/masy.200850210

[18] Akbarzadeh E, Shockravi A, Vatanpour V. High performance compatible thiazole-based polymeric blend cellulose acetate membrane as selective CO2 absorbent and molecular sieve. Carbohydrate Polymers. 2021;**252**:117215. DOI: 10.1016/j.carbpol.2020.117215

[19] Sabir A, Shafiq M, Islam A, Jabeen F, Shafeeq A, Ahmad A, et al. Conjugation of silica nanoparticles with cellulose acetate/polyethylene glycol 300 membrane for reverse osmosis using MgSO4 solution. Carbohydrate Polymers. 2016;**136**:551-559. DOI: 10.1016/j. carbpol.2015.09.042

[20] Li S, Wang D, Xiao H, Zhang H, Cao S, Chen L, et al. Ultra-low pressure cellulose-based nanofiltration membrane fabricated on layer-by-layer assembly for efficient sodium chloride removal. Carbohydrate Polymers. 2021;**255**:117352. DOI: 10.1016/j.carbpol.2020.117352

[21] Lv J, Zhang G, Zhang H, Yang F. Exploration of permeability and antifouling performance on modified cellulose acetate ultrafiltration membrane with cellulose nanocrystals. Carbohydrate Polymers. 2017;**174**:190-199. DOI: 10.1016/j.carbpol.2017.06.064

[22] Zizovic I, Tyrka M, Matyja K, Moric I, Senerovic L, Trusek A. Functional modification of cellulose acetate microfiltration membranes by supercritical solvent impregnation. Molecules. 2021;**26**:411. DOI: 10.3390/molecules26020411

[23] Hassan Abdellatif F, Babin J, Arnal-Herault C, David L, Jonquieres A. Grafting of cellulose acetate with ionic liquids for biofuel purification by a membrane process: Influence of the cation. Carbohydrate Polymers. 2016;**147**:313-322. DOI: 10.1016/j.carbpol.2016.04.008

[24] Ang MBMY, Devanadera KPO, Duena ANR, Luo ZY, Chiao YH, Millare JC, et al. Modifying cellulose acetate mixed-matrix membranes for improved oil-water separation: Comparison between sodium and organo-montmorillonite as particle

additives. Membranes. 2021;**11**(2):80. DOI: 10.3390/membranes11020080

[25] Liu H, Lo HY. Ultrafine fibrous cellulose membranes from electrospinning of cellulose acetate. Journal of Polymer Science: Polymer Physics. 2002;**40**(18):2119-2129. DOI: 10.1002/ polb.10261

[26] Vatankhah E, Prabhakaran MP, Jin G, Mobarakeh LG, Ramakrishna S. Development of nanofibrous cellulose acetate/gelatin skin substitutes for variety wound treatment applications. Journal of Biomaterials Applications. 2014;**28**:909-921. DOI: 10.1177/0885328213486527

[27] Zugenmaier P. Characterization and physical properties of cellulose acetates. Macromolecular Symposia. 2004;**208**:81-166. DOI: 10.1002/masy.200450407

[28] Vatanpour V, Pasaoglu E, Barzegar H, Teber OO, Kaya R, Bastug M, et al. Cellulose acetate in fabrication of polymeric membranes: A review. Chemosphere. 2022;**295**:133-914

[29] Kang F, Singh J. Effect of additives on the release of a model protein from PLGA microspheres. AAPS PharmSciTech. 2001;**2**(2):86-92. DOI: 10.1007/ BF02830570

[30] Haddad R, Ferjani E, Roudesli MS, Deratani A. Properties of cellulose acetate nanofiltration membranes. Application to brackish water desalination. Desalination. 2004;**167**:403-409. DOI: 10.1016/j. desal.2004.06.154

[31] Kutowy O, Sourirajan S. Cellulose acetate ultrafiltration membranes. Journal of Applied Polymer Science. 1975;**19**(5):1449-1460. DOI: 10.1002/ app.1975.070190525

[32] Ye SH, Watanabe J, Iwasaki Y, Ishihara K. Novel cellulose acetate membrane blended with phospholipid polymer for hemocompatible filtration system. Journal of Membrane Science. 2002;**210**:411-421. DOI: 10.1016/ S0376-7388(02)00421-0

[33] Tahazadeh S, Mohammadi T, Tofighy MA, Khanlari S, Karimi H, Motejadded Emrooz HB. Development of cellulose acetate/metal-organic framework derived porous carbon adsorptive membrane for dye removal applications. Journal of Membrane Science. 2021;**638**:119692. DOI: 10.1016/j. memsci.2021.119692

[34] Sharma PR, Sharma SK, Lindström T, Hsiao BS. Nanocellulose-enabled membranes for water purification: Perspectives. Advanced Sustainable Systems. 2020;**4**:1-28. DOI: 10.1002/adsu.201900114

[35] Syamani FA. Cellulose-based membrane for adsorption of dye in batik industry wastewater. International Journal of Hydrology. 2020;**4**:281-283. DOI: 10.15406/ijh.2020.04.00255

[36] Araújo T, Bernardo G, Mendes A. Cellulose-based carbon molecular sieve membranes for gas separation: A review. Molecules. 2020;**25**:1-33. DOI: 10.3390/ molecules25153532

[37] Muhmed SA, Nor NAM, Jaafar J, Ismail AF, Othman MHD, Rahman MA, et al. Emerging chitosan and cellulose green materials for ion exchange membrane fuel cell: A review. Energy, Ecology and Environment. 2020;**5**:85-107. DOI: 10.1007/ s40974-019-00127-4

[38] Abdelhamid AE, Khalil AM. Polymeric membranes based on cellulose acetate loaded with candle soot nanoparticles for water desalination. Journal of Macromolecular Science, Part A: Pure and Applied Chemistry. 2019;**56**:153-161. DOI: 10.1080/10601325.2018.1559698

[39] Idress H, Zaidi SZJ, Sabir A, Shafiq M, Khan RU, Harito C, et al. Cellulose acetate based complexation-NF membranes for the removal of Pb(II) from waste water. Scientific Reports. 2021;**11**:1-14. DOI: 10.1038/s41598-020-80384-0

[40] Sajjan P, Nayak V, Padaki M, Zadorozhnyy VY, Klyamkin SN, Konik PA. Fabrication of cellulose acetate film through blending technique with palladium acetate for hydrogen gas separation. Energy & Fuels. 2020;**34**:11699-11707. DOI: 10.1021/acs.energyfuels.0c02030

[41] De Guzman MR, Andra CKA, Ang MBMY, Dizon GVC, Caparanga AR, Huang SH, et al. Increased performance and antifouling of mixed-matrix membranes of cellulose acetate with hydrophilic nanoparticles of polydopamine sulfobetaine methacrylate for oil-water separation. Journal of Membrane Science. 2021;**620**:118881. DOI: 10.1016/j.memsci.2020.118881

[42] Rasool MA, Van Goethem C, Vankelecom IFJ. Green preparation process using methyl lactate for cellulose-acetate-based nanofiltration membranes. Separation and Purification Technology. 2020;**232**:115903. DOI: DOI.org/10.1016/j.seppur.2019.115903

[43] Wang HH, Jung JT, Kim JF, Kim S, Drioli E, Lee YM. A novel green solvent alternative for polymeric membrane preparation via nonsolvent-induced phase separation (NIPS). Journal of Membrane Science. 2019;**574/**:44-54. DOI: 10.1016/j.memsci.2018.12.051

[44] Rasool MA, Vankelecom IFJ. Preparation of full-bio-based nanofiltration membranes. Journal of Membrane Science. 2021;**618**:118674. DOI: 10.1016/j.memsci.2020.118674

[45] Safarpour M, Safikhani A, Vatanpour V. Polyvinyl chloride-based membranes: A review on fabrication techniques, applications and future perspectives. Separation and Purification Technology. 2021;**279**:119678. DOI: 10.1016/j.seppur.2021.119678

[46] Chen Z, Deng M, Chen Y, He G, Wu M, Wang J. Preparation and performance of cellulose acetate/polyethyleneimine blend microfiltration membranes and their applications. Journal of Membrane Science. 2004;**235**:73-86. DOI: 10.1016/j.memsci.2004.01.024

[47] Saljoughi E, Mohammadi T. Cellulose acetate (CA)/polyvinylpyrrolidone (PVP) blend asymmetric membranes: Preparation, morphology and performance. Desalination. 2009;**249**:850-854. DOI: 10.1016/j.desal:2008:12.066

[48] Sairam M, Sereewatthanawut E, Li K, Bismarck A, Livingston AG. Method for the preparation of cellulose acetate flat sheet composite membranes for forward osmosis-desalination using $MgSO_4$ draw solution. Desalination. 2011;**273**:299-307. DOI: 10.1016/j.desal.2011.01.050

[49] Abedini R, Mousavi SM, Aminzadeh R. A novel cellulose acetate (CA) membrane using TiO2 nanoparticles: Preparation, characterization and permeation study. Desalination. 2011;**277**:40-45. DOI: 10.1016/j.desal.2011.03.089

[50] Ahmad A, Jamshed F, Riaz T, Sabad-E-Gul WS, Sabir A, Alanezi AA, et al. Self-sterilized composite membranes of cellulose acetate/polyethylene glycol for water desalination. Carbohydrate Polymers. 2016;**149**:207-216. DOI: 10.1016/j.carbpol.2016.04.104

[51] Ahmad AL, Jawad ZA, Low SC, Zein SHS. A cellulose acetate/

multi-walled carbon nanotube mixed matrix membrane for CO_2/N_2 separation. Journal of Membrane Science. 2014;**2014**(451):55-66. DOI: 10.1016/j.memsci.2013.09.043

[52] Sossna M, Hollas M, Schaper J, Scheper T. Structural development of asymmetric cellulose acetate microfiltration membranes prepared by a single-layer dry-casting method. Journal of Membrane Science. 2007;**289**:7-14. DOI: 10.1016/j.memsci.2006.11.024

[53] Ounifi I, Saidi N, Kahloul M, Sealey KS, Hafiane A, Ferjani E. Synthesis and characterization of ultrafiltration membranes by phase inversion and by UroPathogenic Escherichia coli retention performance. Desalination and Water Treatment. 2019;**163**:109-117. DOI: 10.5004/dwt.2019.24085

[54] Ounifi I, Ursino C, Santoro S, Chekir J, Hafiane A, Figoli A, et al. Cellulose acetate nanofiltration membranes for cadmium remediation. Journal of Membrane Science and Research. 2020;**6**:226-234. DOI: 10.22079/JMSR.2020.120669.1336

[55] Ounifi I, Guesmi Y, Ursino C, Santoro S, Mahfoudhi S, Figoli A, et al. Antifouling membranes based on cellulose acetate (CA) blended with poly (acrylic acid) for heavy metal remediation. Applied Sciences. 2021;**11**:4354. DOI: 10.3390/app11104354

[56] Ounifi I, Ursino C, Hafiane A, Figoli A, Ferjani E. Preparation of thin film composite membranes using interfacial polymerization for treatment of industrial water containing heavy metals. Desalination and Water Treatment. 2019;**170**:80-90. DOI: 10.5004/dwt.2019.24723

[57] Ounifi I, Guesmi Y, Ursino C, Castro-Muñoz R, Agougui H, Jabli M, et al. Synthesis and characterization of a thin-film composite nanofiltration membrane based on polyamide-cellulose acetate: Application for water purification. Journal of Polymers and the Environment. 2022;**30**(2):707-718. DOI: 10.1007/s10924-021-02233-z

[58] Ounifi I, Guesmi Y, Ursino C, Agougui H, Jabli M, Hafiane A, et al. Synthesis of a thin-film polyamide cellulose acetate membrane: Effect of monomers and porosity on Nano-filtration performance. Journal of Natural Fibers. 2021;**19**:1-18. DOI: 10.1080/15440478.2021.2002766

[59] Pak SH, Jeon YW, Shin MS, Koh HC. Preparation of cellulose acetate hollow-fiber membranes for CO_2/CH_4 separation. Environmental Engineering Science. 2016;**33**(1):17-24. DOI: 10.1089/ees.2015.0201

[60] Pagliero M, Khayet M, García-Payo C, García-FernándezL. Hollow fibre polymeric membranes for desalination by membrane distillation technology: A review of different morphological structures and key strategic improvements. Desalination. 2021;**516**:115235. DOI: 10.1016/j.desal.2021.115235

[61] Su Y, Chen C, Li Y, Li J. PVDF membrane formation via thermally induced phase separation. Journal of Macromolecular Science, Part A: Pure and Applied Chemistry. 2007;**44**:99-104. DOI: 10.1080/10601320601044575

[62] Figoli A. Thermally induced phase separation (TIPS) for membrane preparation. In: Encycl. Membr. Berlin, Heidelberg: Springer Berlin Heidelberg; 2014. pp. 1-2. DOI: 10.1007/978-3-642-40872-4_1866-1

[63] Fan Z, Xiao C, Liu H, Huang Q, Zhao J. Structure design and performance

study on braid-reinforced cellulose acetate hollow fiber membranes. Journal of Membrane Science. 2015;**486**:248-256. DOI: 10.1016/j.memsci.2015.03.066

[64] Chou WL, Yang MC. Effect of take-up speed on physical properties and permeation performance of cellulose acetate hollow fibers. Journal of Membrane Science. 2005;**250**:259-267. DOI: 10.1016/j.memsci.2004.10.030

[65] Dünweg G, Steinfeld L, Ansorge W. Dialysis membrane made of cellulose acetate. 1995; US 5403485

[66] Zhang XL, Xiao CF, Hu XY, Jin X, Bai QQ. Study on the interfacial bonding state and fouling phenomena of polyvinylidene fluoride matrix-reinforced hollow fiber membranes during microfiltration. Desalination. 2013;**330**:49-60. DOI: 10.1016/j.desal.2013.09.022

[67] Su J, Yang Q, Teo JF, Chung TS. Cellulose acetate nanofiltration hollow fiber membranes for forward osmosis processes. Journal of Membrane Science. 2010;**355**(1-2):36-44. DOI: 10.1016/j.memsci.2010.03.003

[68] Bendrea AD, Cianga L, Cianga I. Review paper: Progress in the field of conducting polymers for tissue engineering applications. Journal of Biomaterials Applications. 2011;**261**:3-84. DOI: 10.1177/088532821140270

[69] Lyu Q, Peng B, Xie Z, Du S, Zhang L, Zhu J. Moist-induced electricity generation by electrospun cellulose acetate membranes with optimized porous structures. ACS Applied Materials & Interfaces. 2020;**12**(51):57373-57381. DOI: 10.1021/acsami.0c17931

[70] Cheng J, Zhan C, Wu J, Cui Z, Si J, Wang Q, et al. Highly efficient removal of methylene blue dye from an aqueous solution using cellulose acetate nanofibrous membranes modified by polydopamine. ACS Omega. 2020;**5**(10):5389-5400. DOI: 10.1021/acsomega.9b04425

[71] Xu Q, Peng J, Zhang W, Wang X, Lou T. Electrospun cellulose acetate/P (DMDAAC-AM) nanofibrous membranes for dye adsorption. Journal of Applied Polymer Science. 2020;**137**(15):48565. DOI: 10.1002/app.48565

[72] Li C, Lou T, Yan X, Long Y, Cui G, Wang X. Fabrication of pure chitosan nanofibrous membranes as effective absorbent for dye removal. International Journal of Biological Macromolecules. 2018;**106**:768-774. DOI: 10.1016/j.ijbiomac.2017.08.072

[73] Vatanpour V, Pasaoglu ME, Barzegar H, Teber OO, Kaya R, Bastug M, et al. Cellulose acetate in fabrication of polymeric membranes: A review. Chemosphere. 2022;**295**:133-935

[74] Allen NE, Obendorf SK, Fan J. Polyoxometalate (POM) grafted grooved nanofibrous membranes for improved self-decontamination. RSC Advances. 2016;**6**(89):85985-85993

[75] Asiri AM, Petrosino F, Pugliese V, Khan SB, Alamry KA, Alfifi SY, et al. Synthesis and characterization of blended cellulose acetate membranes. Polymers. 2021;**14**(1):4

[76] Nainar MG, Jayaraman K, Meyyappan HK, Miranda LR. Antifouling properties of poly (vinylidene fluoride)-incorporated cellulose acetate composite ultrafiltration membranes. Korean Journal of Chemical Engineering. 2020;**37**(12):2248-2261

[77] Bolandi S, Ashtiani FZ, Okhovat A, Ghandashtani MB. The

effects of surfactant additives on properties and morphologies of cellulose acetate membranes prepared by phase inversion. Theoretical Foundations of Chemical Engineering. 2020;**54**(5):931-939

[78] Pereira BDS, Moreti LOR, Silva MF, Bergamasco R, Piccioli ADFB, Garcia EE, et al. Water permeability increase in ultrafiltration cellulose acetate membrane containing silver nanoparticles. Materials Research. 2018;**20**:887-891

[79] Mendes G, Faria M, Carvalho A, Gonçalves MC, Pinho MN. Structure of water in hybrid cellulose acetate-silica ultrafiltration membranes and permeation properties. Carbohydrate Polymers. 2018;**189**:342-351

[80] Najafi M, Sadeghi M, Bolverdi A, PourafshariChenar M, Pakizeh M. Gas permeation properties of cellulose acetate/silica nanocomposite membrane. Advances in Polymer Technology. 2018;**37**:2043-2052. DOI: 10.1002/adv.21862

[81] Yang X, Liu Y, Hu S, Yu F, He Z, Zeng G, et al. Construction of Fe3O4@MXene composite nanofiltration membrane for heavy metal ions removal from wastewater. Polymers for Advanced Technologies. 2021;**32**(3):1000-1010. DOI: 10.1002/pat.5148

[82] Gebru KA, Das C. Removal of chromium (VI) ions from aqueous solutions using amine-impregnated TiO_2 nanoparticles modified cellulose acetate membranes. Chemosphere. 2018;**191**:673-684

[83] Vatanpour V, Faghani S, Keyikoglu R, Khataee A. Enhancing the permeability and antifouling properties of cellulose acetate ultrafiltration membrane by incorporation of ZnO@ graphitic carbon nitride nanocomposite. Carbohydrate Polymers. 2021;**256**:117413

[84] Kumar M, Isloor AM, Somasekhara Rao T, Ismail AF, Farnood R, PMG N. Removal of toxic arsenic from aqueous media using polyphenylsulfone/cellulose acetate hollow fiber membranes containing zirconium oxide. Chemical Engineering Journal. 2020;**393**:124367. DOI: 10.1016/j.cej.2020.124367

[85] El-Din LAN, El-Gendi A, Ismail N, Abed KA, Ahmed AI. Evaluation of cellulose acetate membrane with carbon nanotubes additives. Journal of Industrial and Engineering Chemistry. 2015;**26**:259-264. DOI: 10.1016/j.jiec.2014.11.037

[86] Ghaseminezhad SM, Barikani M, Salehirad M. Development of graphene oxide-cellulose acetate nanocomposite reverse osmosis membrane for seawater desalination. Composites Part B:Engineering. 2019;**161**:320-327

[87] Azhar O, Jahan Z, Sher F, MBK N, Kakar SJ, Shahid M. Cellulose acetate-polyvinyl alcohol blend hemodialysis membranes integrated with dialysis performance and high biocompatibility. Materials Science and Engineering: C. 2021;**126**:112127

[88] Doosti M, Abedini R. Polyethyleneglycol-modified cellulose acetate membrane for efficient olefin/paraffin separation. Energy & Fuels. 2022;**36**(17):10082-10095

[89] Riaz T, Ahmad A, Saleemi S, Adrees M, Jamshed F, Hai AM, et al. Synthesis and characterization of polyurethane-cellulose acetate blend membrane for chromium (VI) removal. Carbohydrate Polymers. 2016;**153**:582-591. DOI: 10.1016/j.carbpol.2016.08.011

[90] Diainabo KJ, Mthombeni NH, Motsa M. Preparation and characterization

of hybrids of cellulose acetate membranes blended with polysulfone and embedded with silica for Copper (II), Iron (II) and Zinc (II) removal from contaminated solutions. Journal of Polymers and the Environment. 2021;**29**(11):3587-3604. DOI: 10.1007/s10924-021-02094-6

[91] Azhar AH, Mohamed SH, Thanaa IS, Elkady MF, Abd MA, Elkawi HA. Hamad: Electrospun cellulose acetate nanofiber in corporated with hydroxyapatite for removal of heavy metals. International Journal of Biological Macromolecules. 2019;**151**:1299-1313. DOI: 10.1016/j.ijbiomac.2019.10.176

[92] Reem AW, Ahmed MK, Mansour SF. Tuning the synthetic conditions of grapheneoxide/magnetite/hydroxyapatite/cellulose acetate nanofibrous membranes for removing Cr(VI), Se(IV) and methylene blue from aqueous solutions. Journal of Water Process Engineering. 2020;**38**:101543. DOI: 10.1016/j.jwpe.2020.101543

[93] Cheng J, Zhan C, Wu J, Cui Z, Junhui S, Wang Q, et al. Highly efficient removal of methylene blue dye from an aqueous solution using cellulose acetate nanofibrous membranes modified by polydopamine. American Chemical Society. 2020;**5**:5389-5400. DOI: 10.1021/acsomega.9b04425

[94] Shirin T, Toraj M, Maryam AT, Samaneh K, Hamid K, BME H. Development of cellulose acetate/metal-organicframeworkderivedporouscarbo nadsorptive membrane for dyeremoval applications. Journal of Membrane Science. 2021;**638**:119692. DOI: 10.1016/j.memsci.2021.119692

[95] Muna AD, Saja AR, Borhan AA. Photo catalytic degradation of methylene blue using polymeric membranes based on cellulose acetate impregnated with ZnO nanostructures. Polymers. 2021;**13**:3451. DOI: 10.3390/polym13193451

[96] Mohd Amin NAA, Mokhter MA, Salamun N, Wan MAWM. Phosphate adsorption from aqueous solution using electrospun cellulose acetate nanofiber membrane modified with graphene oxide/sodium dodecyl sulphate. Membranes. 2021;**11**:546. DOI: 10.3390/membranes11070546

[97] José L, Ramírez C, Xaimara SM, Simara L, Pedro OMF, Marielys TD, et al. Porous cellulose acetate/block copolymer membranes for the recovery of polyphenolic compounds from aquatic environments. ACS Omega. 2022;7:2774-2785. DOI: 10.1021/acsomega.1c04327

[98] Mohit C, Abhijit M. Fe–Al–Mn@ chitosan based metaloxides blended cellulose acetate mixed matrix membrane for fluoridede contamination from water: Removal mechanisms and antibacterial behavior. Journal of Membrane Science. 2020;**611**:118-372. DOI: 10.1016/j.memsci.2020

[99] Ye T, Min W, Ruigang L, Yanxiang L, Deqian W, Junjun T, et al. Electrospun membrane of cellulose acetate for heavy metal ion adsorption in water treatment. Carbohydrate Polymers. 2011;**83**:743-748. DOI: 10.1016/j.carbpol.2010.08.054

[100] Damiri F, Andra S, Kommineni N, Balu SK, Bulusu R, Boseila AA, et al. Recent advances in adsorptive nanocomposite membranes for heavy metals ion removal from contaminated water: A comprehensive review. Materials. 2022;**15**(15):5392. DOI: 10.3390/ma15155392

Chapter 3

Conversion of Cellulose into Cellulose Acetate and Evaluation of Biomedical and Wastewater Cleaner Application of Electrospun Cellulose Acetate Nanofibers

Rebika Baruah and Archana Moni Das

Abstract

Conversion of biomass into useful organic chemicals has great demand in the twenty-first century. Cellulose is the most natural abundant biopolymer. Production of cellulose acetate (CA) from cellulose has garnered tremendous global attention because of their myriad application. CA is produced in huge amounts globally. Annual production of the derivative crossed 1.5 billion pounds. CA has remarkable biomedical applications due to their capability of drug delivery and anti-inflammatory, anti-cancer, antioxidant, antimicrobial properties. This chapter highlighted the synthesis and chemical and physical properties of CA polymer and electrospun CA nanofibers (CA NF) and their tremendous applications in biomedical and wastewater treatment.

Keywords: cellulose, cellulose acetate, nanofibers, biomedical, wastewater treatment

1. Introduction

Electrospinning is one of the cost-effective procedures that were applied to synthesize multifunctional nanofiber from the origin of this century [1, 2]. Since the mid-1990s, researchers started to explore the big potential of nanofiber in the area of nanotechnology [3]. Die to the large surface area to volume ratio, about a thousand times higher than that of a human hair, nanofiber are tremendously applicable engineering material. Nanofibers (NFs) have diverse application in various fields such as catalysis, tissue scaffolds [4, 5], textiles, wastewater remediation, drug delivery, wound healing, food packaging, and electronics [6]. NFs are synthesized via electrospinning is grounded on the uniaxial unfolding of a visco-elastic mixture. The precepts of electrospinning and different factors like time, temperature, etc., affects the shape of the nanofiber assembly. Different techniques of electrospinning such as dry-spinning and melt-spinning build develop electrostatic forces to stretch the solution as it solidifies. In the electrospinning the drawing of the solution to make the fiber will proceed as long as there is enough solution to feed the electrospinning jet. This is the advantages of electrospinning over the formal fiber spinning techniques. Synthesis of fiber is continuous without any interruption to the electrospinning jet [7].

Researchers of this era search out for economically feasible and commercially successful pathways to convert biomass into useful organic chemicals [8]. Cellulose is the most abundant natural biopolymer in the world. Cellulose is converted into different types of valuable cellulosic derivatives. Among derivatives, cellulose acetate gathers the attention of researchers due to its cost-effective, sustainable and eco-friendly nature [9]. According to the survey in 2005, the annual production of cellulose acetate is 1.5 billion pounds due its versatile applications in different fields [10]. Traditionally, cellulose is reacted with acetic acid and acetic anhydride in presence of sulfuric acid and synthesized cellulose acetate. Now days, some innovative and effective methods were developed to synthesize cellulose acetate, e.g., use of ionic liquids, green catalyst like nanoparticles, super acids like SO_4^{2-}/ZrO_2 in a solvent free ball milling reactor, etc. [11].

Cellulose acetate is applied as coating materials in textile industry, waste water treatment for filtration, pharmaceutical industry, stabilization of nanoparticles, paper industry, leather, dye, and food industries, and lamination, etc. [12]. Cellulose acetate/montmorillonite bionanocomposites were applied in the adsorption of organic dyes like Methylene blue, Methyl Orange, Orange G, Rhodamine B, Eosin Y, etc. They were degraded the harmful dyes by adsorption due to their large surface to volume ratio [13, 14]. Cellulose acetate based nanocomposites can be effectively used in the remediation of organic pollutants and safe the world from the detrimental effects of water pollutants. They can successfully overcome the disadvantages of traditional method used in waste water treatment like ultra-filtration, electro-chemical adsorption, coagulation, photooxidation, ultrasound-assisted, flocculation-coagulation, etc. Bionanocomposites effectively remediates the pollutants through physical adsorption. Physical adsorption is an inexpensive, highly efficient method to remove harmful entities in waste water and make the water safe for drinking [15].

Cellulose acetate (CA) based bionanocomposites have profound biomedical applications, like coating of medicine, membrane, drug delivery, blood purification in chronic renal dysfunction, hemodialysis, etc. Their biodegradability, renewability, non-corrosivity, non-toxicity, sustainability, and biocompatibility make them a suitable and innovative alternative in the place of multi-drug resistant medicines [16]. As dialyzers cellulose and its derivatives are 1st generation of polymers, they play an important role in the removal of blood toxins by diffusion and convection. This chapter is designed to explore the wastewater remediation capability and biomedical applicability of cellulose acetate based bionanocomposites.

2. Cellulose acetate (CA) electrospun nanofibers

Cellulose acetate can be synthesized from cellulose via a green and economic method and it has tremendous application in food technology for packaging and biomedical fields as CA is an easily obtainable substantial derived from cellulose at low cost and it is for regenerative medicine, tissue engineering, bone regeneration, and stem cell research [17, 18]. CA is more easily electrospinned as compared to cellulose. Conversion of nanocellulose from cellulose acetate is easier than that from cellulose. Electrospinned cellulose acetate nanofiber (CA NF) is completely novel candidate in the field of biomedical and food industry. Global industrial demand of Ca fiber is about 1.05 M metric tons by 2017 [19, 20]. Environment friendly CA NF has great demand in the industrial world due to their biodegradability, higher tensile strength, enough angularity, and more modulus [21–23]. Electrospun cellulose acetate nanofibers have commendable applications in different fields and introduction of new functional groups into the CA NF increases the utilization of this NF to greater extent [24, 25]. CA NF possesses many beneficial characteristics like biocompatible, biodegradable, water resistance, etc. Therefore CANF can be safely used in the preparation of bandage loaded with nanocomposites for biomedical uses [26].

2.1 Choice of solvent for cellulose acetate e-spun nanofabrication

Choice of solvent is an important step to synthesize electrospun CA NF. It affects the morphology, size, and shape of the NF [27]. Researchers used different types of solvents like chloroform, N, N-dimethylformamide, dichloromethane, formic acid, and methanol (MeOH) and pyridine and they also used mixture of solvents like acetone, dimethyl acetamide (DMAc), chloroform-MeOH, and DCM-MeOH [28]. Formhals synthesized electrospun CA NF in acetone as the solvent [29]. Viscosity, surface tension, and conductibility of solvent system play a vital role in the development of smooth NF. Acetone, DMF, trifluro-ethanol and mixture of acetone and water are the best solvent systems for preparation of e-spinning CA NF [30, 31]. CA NF with poly(ethyleneglycol), poly(ethylene oxide) and hydroxyapatite becomes an effective bionanocomposites in the mentioned solvent systems [32]. Haas et al. reported the effect of solvent system in the restraining of the coordination between CA NF [33]. Binary solvent system of less volatile solvents produced ribbon like structure with good packaging property [34]. Humidity and temperature of solvents have adverse effects on the shape and size of CA NF [34]. Along with solvent system, the field strength, distance between tip to collector, feed rate of solution affects the properties of CA NF like structure, diameter of fiber, etc. [35]. By using Box-Behnken pattern method the affected property of the CA NF can be modified [36]. Researchers concern for the preparation of e-spinning CA NF by blending with natural polymer and metal nanoparticles like AgNPs, etc. [37].

3. Properties of cellulose acetate

3.1 Thermal properties of cellulose acetate

Thermal properties like glass transition temperature (Tg), melting point (Tm) and thermal decomposition (Td) of CA is dependent on degree of substitution (DS) and average molecular weight [38]. Plasticizers are used with CA to prepare thermoplastic to overcome the disadvantages of decomposition [39, 40]. Liu et al. studied the effects of degree of substitution on the thermal properties of CA NF in the aqueous hydrolysis of CA. He stated that DS have negative effects on Tg. When DS deceased from 2.45 to 1.77 Tg increased from 198°C to 205°C [41]. Elevation of Tg was due to the increasing number of free hydroxyl groups at low DS. On mixing of CA with copolymers nanofiber also increases the Tg [42]. Intermolecular hydrogen bonds and dipole-dipole interactions between the monomers of CA polymer have significant impact on the properties of CA NF. Number of both of hydrogen and acetyl groups affects the density and strength of hydrogen bond and dipole-dipole interactions of CA [41]. FTIR spectra of CA confirmed the presence of hydroxyl (OH) stretching band at 3322 cm^{-1} and the carbonyl (CO) stretching band at 1751 cm^{-1}. On increasing degree of acetylation, the reduction of hydroxyl groups and increasing of carbonyl group were achieved [41]. Electrospun CA NF that possessed DS value of 2.4 or 2.5 can be effectively utilized in the biomedical application [41]. Inner structure of CA NF is more stable than original CA. CA NF possessed higher thermal stability and broad range of thermal degradation as compared to CA [42]. Presence of acetyl groups in presence of hydroxyl groups in the CA NF is the reason for the higher thermal stability of NF [42].

3.2 Mechanical properties of electrospun cellulose acetate

Mechanical properties of CANF are analyzed by measuring the tensile strength, young modulus, and elasticity of fiber. Mechanical properties are dependent on the morphological structure and packing of nano fiber. Mechanical properties help to determine the suitability of application of CA NF in biomedical fields [43, 44]. Golizadeh et al. reported the tensile strength and Young's modulus of CA NF approximately 2.5 MPa and nearly 125 MPa [44]. In their study the tensile strength and Young's modulus were increased by chemical and heating treatment of Ca NF. Bonding of CANF at crossover points occurred on heating due to the cross linking of nanofibers and it increased the tensile strength of CANF [44].

Van der Waals interaction and fiber-fiber fusion in CANF also affect the mechanical properties of NF [45]. Increase in the cross-linking between fiber improves the mechanical properties of fiber like, shear resistance, tensile strength, and compressive [46]. Mixing of another polymer with CANF also improves the mechanical strength of nano fiber. The mechanical properties of CA could be enhanced by mixing with another polymer. E. g., Poly hydroxybutyrate (PHB)/CANF scaffold possessed high tensile strength and Young's modulus at 7.86 ± 0.67 MPa and 854.2 ± 187.6 MPa whereas the pure CANF has low tensile strength and Young's modulus was 1.56 ± 0.19 MPa and 41.6 ± 12.3 MPa, respectively. Improvement of mechanical properties of CANF was due to the intermolecular hydrogen bond in CANF and intermolecular interactions of C=O groups in PHB [46]. Increase in the bonding of CANF and adjustment in the degree of molecular orientation of CA also improves the mechanical properties of nano fiber [46].

3.3 Biodegradability of cellulose acetate

Cellulose acetate nano fiber is a biodegradable polymer. Microorganisms degrade CANF in both aerobic and anaerobic condition by using cellulase and acetyl esterase enzymes [20]. Degree of polymerization, degree of substitution, and location affect the rate of degradation of CANF [21]. Acetyl-esterase is an essential enzyme for the initial step of biodegradation of CANF and cellulase is important in second step in which hydrolysis of the backbone of cellulose take place [22]. In the biodegradation, acetyl group from the C2- and C3-positions of the CA convert to acetic acid in first step and in hydrolysis free glycosyl residues release into the metabolic cycle [21]. Puls et al. reported the biodegradation of CANF with the help of cellobiohydrolase I (CBHI) and *cellobiohydrolase* II (CBHII) as cellulolytic enzymes. They catalyze the splitting of β-1,4-glycosidic bonds at the reducing and non-reducing end side of the cellulose chain.

CANF from bacterial cellulose loaded with cellulose is degraded by the *in vitro* degradation. *In vitro* biodegradation converts CANF to simulated body fluid within 6 months. Modulation of the content of cellulose modifies the rate of degradation [44].

Although many studies about the biodegradation CANF are done in microorganisms but biodegradation of CANF has not been studied yet. Therefore insertion of CANF loaded composites in to the human body is a challengeable research till now.

3.4 Biocompatibility of cellulose acetate

Biocompatibility is defined as the behavior of implanted biomaterials in living tissues without causing any local or systemic adverse effect in their surroundings. Biomaterials also possess an effective host response in a specific application [46].

Biocompatibility test is carried out under the guidance of the International Organization for Standardization (ISO) 10993 series. The aim of this test to protects the humans from future biological risks of medical devices or materials [47]. Positive results of this test provide the final approval of implantation of biomaterials in humans. Biocompatibility tests include cytotoxicity, sensitization, irritation, systemic toxicity, genotoxicity, and hemocompatibility tests [47]. CANF possessed excellent biocompatibility [48]. Liang et al. reported the biocompatibility of poly (ethylene terephthalate) (CDA-g-PET) grafted CANF and showed that the viability of fibroblast on CDA-g-PET grafted CANF significant increase with incubation time.

Biocompatibility of other polymers can be improved by addition of CA to them. E. g., electrospun CANF grafted PHB showed better biocompatibility than pure HPOB film [48]. Biocompatibility of CANF/carboxymethylcellulose (CMC) scaffolds showed better results than that of pure CMC scaffolds [48].

4. Application of cellulose acetate nanofiber

4.1 Waste water treatment

More than one billion people in the world are suffered from lack of safe drinking water [49]. Remediation of waste water is a great challenge for researchers till now [49]. Many industries like textiles, dye, pharmaceutical, etc. discharged their waste water to water bodies and cause water pollution. Water pollutants like, harmful dyes, heavy metals such as, Fe, etc., and pathogens have detrimental effects on the aquatic lives and terrestrials organisms [50]. Traditional methods like chemical precipitation, membrane separation, electrocoagulation, anodic oxidation, TiO_2 photocatalysis, biodegradation, adsorption, and ion exchange are applied in the purification of waste water. Adsorption is the effective one that overcomes the all the disadvantages belongs to other methods and it possesses easily handling, reversibility, recyclability and cost-effective nature [51].

Utilization of bio-based products in the adsorption of organic pollutant in the remediation of waste water is a fruitful idea due to their natural availability, biodegradability, sustainable, and biocompatibility [52]. Cellulose is the most preferred candidate in this category but the production of submicron cellulose fibers is difficult due to the insolubility of cellulose in conventional solvents [53]. To overcome these obstacles cellulose derivatives are synthesized to access their applicability in different fields. Cellulose acetate is the most preferable one due to its environment friendly and economic nature. CA possesses versatile application in many industries like textiles, textiles, cigarette filters, surface coating, inks, motion picture film, microfilm, and audio tape as a carrier [54]. CA is soluble in non-polar solvents and therefore it can be easily electrospinned to from CANF membrane with high mechanical strength [55].

CANF with ribbon-like porous structure successfully prepared to adsorb Cu^{2+} ions were studied. CANF could remove the Cu^{2+} ions up to 88.6% [55].

Fabrication of metal nanoparticles and another biopolymer on the CANF as inorganic filler overcome the drawback of lack of active site on the Ca backbone [56]. CA acts as soft matrix in the accommodation of inorganic fillers on it to functionalize it. E.g., chitosan/CANF composites effectively removed copper from waste water [57]. CANF/HAp bionanocomposites successfully removed harmful bisphenol [58]. These bionanocomposites have size of 92 nm and a new chromatographic method was developed for the quantification of bisphenol A in samples of baby food [58].

4.2 Biomedical application

Incorporation of inorganic nanofillers with CANF makes the bionanocomposites tremendous applicant in the field of biomedical. Due the biodegradability, renewability, non-corrosiveness, non-toxicity and biocompatibility nature cellulose acetate based bionanocomposites have tremendous applications in medicinal fields, like coating of medicine, drug delivery, tissue engineering, bandage preparation, etc. [59].

4.2.1 Antimicrobial activity

Infectious diseases are become threatened to various industries like textile, food, health care and packaging industry. Due to the uses of limitless antibiotics without proper knowledge, microorganism become multi-drug resistant and cause persistent infections. Developing a sterile and cost-effective medicine is an urgent need for twenty-first century that can control the infection caused by multi-drug resistant microbes. This kind of clinically approved medicine will decrease the infection of microorganism in medical instruments, filters, implants, tissue scaffolds, wound dressing materials, etc. [59]. Inorganic antimicrobial nanoparticles like Ag, Cu, and ZnO, and organic materials like N-halamine and chlorhexidine can be incorporated to CANF and increased the antimicrobial property of CANF [59]. Types of incorporated polymers, shape, size, dispersion, and load ratio of nanoparticles adversely affects the antimicrobial activities of CANF based bionanocomposites. Therefore, appropriate inorganic fillers should be incorporated to CANF which have positive effects on the biomedical activity of CANF materials [60].

4.2.2 Packing biomedical and pharmaceutical products

Contamination of food and pharmaceuticals by surroundings is very common and causes undesirable side effects. Extra care is required in the packaging of these products to safe them microbial contamination. de Moraes et al. developed grapheme oxide (GO)/CA packaging film for pharmaceutical, biomedical, and food products. GO/CA films possessed UV light absorption potential, visible light transferable and non-catalytic property towards polymer degradation [61]. CANF based packaging materials provide homogenous and smooth film [61].

4.2.3 Controlled drug release

Drug delivery systems are developed in such way that they can deliver drug at a specific target, i.e., either tissues or intracellular sites within short period of time. Delivering drug to the site action has advantages of control delivery of drugs, requirement of low concentration of drugs, reduction of dose of drugs, effective accumulation of drug to the location of action and overcome the disadvantages of side effects, instance decomposition of drugs, etc. Different materials are employed in the drug delivery like clay, biopolymer, LDH, etc. Cellulose based drug delivery systems are very effective due to their natural abundance, renewability, biocompatibility, biodegradability, and unique biological activities. Madaeni et al. reported the drug delivery ability of polyethylene glycol (PEG)/CANF bionanocomposites. The composites showed fast rate of drug deliver with zero order kinetics. Hydrophilicity of the composite membrane minimized the agglomeration of the nanocomposites and addition of PEG improved the structure and porosity of composite membrane [62].

4.2.3.1 Non-steroidal anti-inflammatory and antioxidant drug

Inflammation and swelling symptoms are decreased by the use of non-steroidal anti-inflammatory medicines (NSAIDs). NSAIDs have some side effects like gastrointestinal problems due to their poor solubility in the water. To overcome this problem local or targeted delivery of NSAIDs are developed by using electrospun biopolymer coated drug delivery systems [63]. CANF can effectively deliver NSAIDs like naproxen, indomethacin, ibuprofen, and sulindac, etc. CANF transdermally deliver drugs like naproxen with sustainable release [64].

4.2.3.2 Anticancer and antioxidant agents

Electrospun CANF/polyethylene oxide (PEO) effectively delivered potential anticancer drugs like cisplatin. Incorporation of cisplatin in CANF/PEO formed the drug formulation with the repeating dumbbell-shaped structures [65]. Electrospun CANF is utilized in the local delivery of silymarin through the skin in various liver diseases. Silymarin was well incorporated in CANF and released gradually to the targeted site within 20 days [66].

4.2.3.3 Antimicrobial agents

Antimicrobial drugs are incorporated with electrospun CANF and utilized to control the growth of microbes [67]. CANF/poly (vinylpyrrolidone) (PVP) was incorporated with amoxicillin and formulates as (CA/PVP-Amoxi/CA) [68]. Tetracycline was also incorporated with CANF and applied in periodontal reconstitution as an adjuvant.

CANF membrane acts as a prospective degradable drug release device in the drug release study [69]. CANF was incorporated with Bis *N-Chloro*-2,2,6,6-tetramethyl-4-piperidyl sebacate (CI-BTMP) and applied in the controlling of bacterial infection [70]. Ag/TiO_2/CANF was applied as an antibacterial for future bone tissue regeneration. Ag/TiO_2/CANF possessed potential antibacterial activity against gram-negative bacteria (*E. coli* and *P. aeruginosa*) and gram-positive bacteria (*S. aureus*). CANF composites showed significantly improved cell viability and the proliferation of fibroblasts [71, 72].

Khan et al. reported the delivery of silver sulfadiazine (SSD), a topical wound dressing antibiotic via Ag NPs/CANF bionanocomposites [73]. Ag NPs/CANF bionanocomposites possessed remarkable antibacterial properties against gram-negative *E. coli*.

4.2.3.4 Wound dressing

Development and designing of wound dressing materials are vital keeping in mind wound pathogenesis. Wound dressing materials should be non-toxic, antimicrobial, maintain moisture, non-adherent, wound care, prevent bacterial infection, and endorse wound healing. Recently, different biological materials have been used for wound dressing. Among them, polymers that considered as soft biomaterials; such as cellulose derivatives and their nanocomposites have been widely used as wound dressing materials [74].

4.2.3.5 Bone repairing and tissue engineering

In the field of bone repairing and tissue engineering, it is most viable to choose a biocompatible and degradable material that lack of cytotoxic effects and could be replaced by the regenerated tissue biomaterial. In addition, considerable attention should be given to provide a sufficient environment for tissue to promote cell integration, differentiation, and proliferation. All these could be achieved by implantation of the appropriate three-dimensional porous scaffolds seeded by the bioactive molecules [75]. Accordingly, due to the fact that cellulose acetate was not only naturally abundant but also has good hydrolytic stability, good mechanical properties, excellent compatibility, low toxicity and relatively low cost [76], many attempts have been conducted to investigate the possibility of fabricating CANC in tissue and bone engineering [77].

5. Conclusion and future prospective

Cellulose acetate nanofiber can by successfully synthesized via electro spinning. CANF possesses remarkable waste water remediation and biomedical properties. CANF shows potential drug delivery application as locally also transdermally. Biodegradability, biocompatibility, and sustainability of electrospun CANF will be cost effective green material in environmental remediation and biomedical application. Due to the lack of cellulose enzyme in human body, biodegradability of CANF is a challengeable topic till now. Incorporation of drugs in e-spinning CANF with degree of substitution of 2.5° is an effective localized drug delivery system without the risk of second surgery. Excellent solubility, hydrophobic nature, mechanical strength, and thermal property are due to the presence of acetyl groups present in the backbone chain of CA and cross linking between the fibers. CA with 2.5° of substitution has commonly been used in the drug delivery systems due to inherited properties. Therefore, CANF electrospun bionanocomposites will be an eco-friendly and cost effective materials in waste water treatment and localized drug delivery systems. These purposeful properties of CANF based bionanocomposites make them strong materials for industrial applications to provide human clean and safe drinking water and effective drug delivery systems in wound dressing, cancer treatment, liver disease, etc., without the risk of secondary infection [78–80].

Acknowledgements

The authors sincerely acknowledged the Director of CSIR-North East Institute of Science and Technology, Jorhat, Assam for his permission to perform our research with excellent facilities and precious guidance. R.B. acknowledged the Department of Science and Technology, New Delhi, for the Inspire Fellowship (GAP-0754).

Conflict of interest

The authors declare no conflict of interest.

Author details

Rebika Baruah[1,2]* and Archana Moni Das[1,2]

1 Natural Products Chemistry Group, Chemical Science and Technology Division, CSIR-North East Institute of Science and Technology, Jorhat, Assam, India

2 Academy of Scientific and Innovative Research (AcSIR), Ghaziabad, India

*Address all correspondence to: baruahrebika9@gmail.com

References

[1] Cooley JF. Apparatus for Electrically Dispersing Fluids. Google Patents. 1902

[2] Morton WJ. Method of Dispersing Fluids. Google Patents. 1902

[3] Doshi J, Reneker DH. Electrospinning process and applications of electrospun fibers. Journal of Electrostatics. 1995;**35**:151-160

[4] Li WJ, Laurencin CT, Caterson EJ, Tuan RS, Ko FK. Electrospun nanofibrous structure, a novel scaffold for tissue engineering. Journal of Biomedical Materials Research. 2002;**60**:613-621

[5] Wang YK, Yong T, Ramakrishna S. Nanofibres and their influence on cells for tissue regeneration. Australian Journal of Chemistry. 2005;**58**:704-712

[6] Ramakrishna S. An Introduction to Electrospinning and Nanofibers. Singapore: World Scient; 2005

[7] Deitzel J, Kleinmeyer J, Hirvonen J, Tan NB. Controlled deposition of electrospun poly (ethylene oxide) fibers. Polymer. 2001;**42**:8163-8170

[8] Fang X, Reneker D. DNA fibers by electrospinning. Journal of Macromolecular Science, Physics. 1997;**36**:169-173

[9] Baker SC, Atkin N, Gunning PA, Granville N, Wilson K, Wilson D, et al. Characterisation of electrospun polystyrene scaffolds for three dimensional in vitro biological studies. Biomaterials. 2006;**27**:3136-3146

[10] Chew SY, Hufnagel TC, Lim CT, Leong KW. Mechanical properties of single electrospun drug-encapsulated nanofibres. Nanotechnology. 2006;**17**:3880

[11] Kaleemullah K, Ming S. Electrospun cellulose acetate nanofibers and Au@AgNPs for antimicrobial activity—A mini review. Nanotechnology Reviews. 2019;**8**:246-257

[12] Das AM, Ali AA, Hazarika MP. Synthesis and characterization of cellulose acetate from rice husk: Eco-friendly condition. Carbohydrate Polymers. 2014;**112**:342-349

[13] Dufresne A. Nanocellulose :A new ageless bionanomaterial. Materials Today. 2015;**16**:220-227

[14] El Nemr A, Ragab S, El Sikaily A. Testing zinc chloride as a new catalyst for direct synthesis of cellulose di- and tri-acetate in a solvent free system under microwave irradiation. Carbohydrate Polymers. 2016;**151**:1058-1067

[15] Goswami M, Das AM. Synthesis of cellulose impregnated copper nanoparticles as an efficient heterogeneous catalyst for C-N coupling reactions under mild conditions. Carbohydrate Polymers. 2018;**195**:189-198

[16] Goswami M, Baruah D, Das AM. Green synthesis of silver nanoparticles supported on cellulose and their catalytic application in the scavenging of organic dyes. New Journal of Chemistry. 2018;**42**:10868-10878

[17] Harrar HF, Dairi N. Elaboration of cellulose acetate nanobiocomposites using acidified gelatin-montmorillonite as nanofiller: Morphology, properties and biodegradation studies. Polymer Composites. 2018;**34**:1515-1524

[18] Hu J, Yu H, Dai W, Yan X, Hu X, Huang H. Enhanced adsorptive removal of hazardous anionic dye "Congo red"

by a Ni/Cu mixed-component metal–organic porous material. RSC Advances. 2014;**4**:35124-35130

[19] Ahmad A, Setapar SHM, Chuong CS, Khatoon A, Wani WA, Kumar R, et al. Recent advances in new generation dye removal technologies: Novel search for approaches to reprocess wastewater. RSC Advances. 2015;**5**:30801-30818

[20] Aksu Z, Donmez G. A comparative study on the biosorption characteristics of some yeast for Remazol Blue reactive dye. Chemosphere. 2003;**50**:1075-1083

[21] Aksu Z, Karabayır G. Comparison of biosorption properties of different kinds of fungi for the removal of Gryfalan Black RL metal-complex dye. Bioresources Technology. 2003;**99**:7730-7741

[22] Amarsekara AS, Ebede CC. Zinc chloride mediated degradation of cellulose at 200°C and identification of the products. Bioresource Technology. 2009;**100**(21):1-10

[23] Aruguz AZ, Gulen J, Evers RH. Adsorption of methylene blue from aqueos solution on pyrolyzed petrified sediment. Bioresources Technology. 2008;**99**:1503

[24] Asfaram A, Ghaedi M, Agarwal S, Tyagi I, Gupta VK. Removal of basic dye Auramine-O by ZnS:Cu nanoparticles loaded on activated carbon: Optimization of parameters, using response surface methodology with central composite design. RSC Advances. 2015;**5**:18438-18450

[25] Goswami M, Das AM. Synthesis and characterization of a biodegradable cellulose acetate-montmorillonite composite for effective adsorption of Eosin Y. Carbohydrate Polymers. 2019;**206**:863-872

[26] Kurt P, Wood L, Ohman DE, Wynne KJ. Highly effective contact antimicrobial surfaces via polymer surface modifiers. Langmuir. 2007;**23**:4719-4723

[27] Edgar KJ. Cellulose Esters, Organic, Encyclopedia of Polymer Science and Technology. Hoboken, New Jersey: Wiley; 2002

[28] Fischer S, Thümmler K, Volkert B, Hettrich K, Schmidt I, Fischer K. Properties and applications of cellulose acetate. Macromolecular Symposia. 2008;**262**:89-96

[29] Puls J, Wilson SA, Hölter D. Degradation of cellulose acetatebased materials, a review. Journal of Polymers and the Environment. 2011;**19**:152-165

[30] Sousa M, Bras AR, Veiga HIM, Ferreira FC, de Pinho MN, Correia NT, et al. Dynamical characterization of a cellulose acetate polysaccharide. The Journal of Physical Chemistry. 2010;**114**:10939-10953

[31] Aoki D, Teramoto Y, Nishio Y. SH-containing cellulose acetate derivatives, preparation and characterization as a shape memory-recovery material. Biomacromolecules. 2007;**8**:3749-3757

[32] Liu C, Bai R. Adsorptive removal of copper ions with highly porous chitosan/cellulose acetate blend hollow fiber membranes. Journal of Membrane Science. 2006;**284**:313-322

[33] Baharifar H, Honarvarfard E, Haji Malek-kheili M, Maleki H, Barkhi M, Ghasemzadeh A, et al. The potentials and applications of cellulose acetate in biosensor technology. Nanomedicine Research Journal. 2017;**2**:216-223

[34] Samadian H, Salehi M, Farzamfar S, Vaez A, Ehterami A, Sahrapeyma H,

et al. In vitro and in vivo evaluation of electrospun cellulose acetate/gelatin/hydroxyapatite nanocomposite mats for wound dressing applications. Artificial Cells, Nanomedicine, and Biotechnology. 2018;**46**:1-11

[35] Boroumand S, Hosseini S, Salehi M, Faridi MR. Drugloaded electrospun nanofibrous sheets as barriers against postsurgical adhesions in mice model. Nanomedicine Research Journal. 2017;**2**:64-72

[36] Farzamfar S, Naseri-Nosar M, Vaez A, Esmaeilpour F, Ehterami A, Sahrapeyma H, et al. Neural tissue regeneration by a gabapentin-loaded cellulose acetate/gelatin wet-electrospun scaffold. Cellulose. 2018;**25**:1229-1238

[37] Gheibi A, Khoshnevisan K, Ketabchi N, Derakhshan MA, Babadi AA. Application of electrospun nanofibrous PHBV scaffold in neural graft and regeneration, a mini-review. Nanomedicine Research Journal. 2016;**1**:107-111

[38] Khalf A, Madihally SV. Recent advances in multiaxial electrospinning for drug delivery. European Journal of Pharmaceutics and Biopharmaceutics. 2017;**112**:1-17

[39] Li W, Li X, Wang T, Li X, Pan S, Deng H. Nanofibrous mats layer-by-layer assembled via electrospun cellulose acetate and electrosprayed chitosan for cell culture. European Polymer Journal. 2012;**48**:1846-1853

[40] Sadri M, Mohammadi A, Hosseini H. Drug release rate and kinetic investigation of composite polymeric nanofibers. Nanomedicine Research Journal. 2016;**1**:112-121

[41] Samadian H, Mobasheri H, Hasanpour S, Faridi MR. Electrospinning of polyacrylonitrile nanofibers and simulation of electric field via finite element method. Nanomedicine Research Journal. 2017;**2**:87-92

[42] Samadian H, Mobasheri H, Hasanpour S, Majid RF. Needleless Electrospinning System, An Efficient Platform To Fabricate Carbon Nanofibers. Switzerland: Trans Tech Publications; 2017. pp. 78-89

[43] Sperling LE, Reis KP, Pranke P, Wendorff JH. Advantages and challenges offered by biofunctional core–shell fiber systems for tissue engineering and drug delivery. Drug Discovery Today. 2016;**21**:1243-1256

[44] Konwarh R, Karak N, Misra M. Electrospun cellulose acetate nanofibers, the present status and gamut of biotechnological applications. Biotechnology Advances. 2013;**31**:421-437

[45] Li W, Li T, Li G, An L, Li F, Zhang Z. Electrospun H4SiW12O40/cellulose acetate composite nanofibrous membrane for photocatalytic degradation of tetracycline and methyl orange with different mechanism. Carbohydrate Polymers. 2017;**168**:153-162

[46] Formhals A. Process and Apparatus for Preparing Artificial Threads. US Patent No 1,975,504. 1934

[47] Haas D, Heinrich S, Greil P. Solvent control of cellulose acetate nanofibre felt structure produced by electrospinning. Journal of Materials Science. 2010;**45**:1299-1306

[48] Tungprapa S, Puangparn T, Weerasombut M, Jangchud I, Fakum P, Semongkhol S, et al. Electrospun cellulose acetate fibers, effect of solvent system on morphology and fiber diameter. Cellulose. 2007;**14**:563-575

[49] Liu H, Hsieh YL. Ultrafine fibrous cellulose membranes from electrospinning of cellulose acetate. Journal of Polymer Science. 2002;**40**:2119-2129

[50] Liu H, Tang C. Electrospinning of cellulose acetate in solvent mixture N, N-dimethylacetamide (DMAc)/acetone. Polymer Journal. 2007;**39**:65

[51] Ma Z, Kotaki M, Ramakrishna S. Electrospun cellulose nanofiber as affinity membrane. Journal of Membrane Science. 2005;**265**:115-123

[52] Qu J, Meng X, Hu Q, You H. A novel two-stage evaluation system based on a Group-G1 approach to identify appropriate emergency treatment technology schemes in sudden water source pollution accidents. Environmental Science and Pollution Research. 2016;**23**:2789-2801

[53] Elkady M, Shokry H, Hamad H. Effect of superparamagnetic nanoparticles on the physicochemical properties of nano hydroxyapatite for ground water treatment: Adsorption mechanism of Fe (II) and Mn (II). RSC Advances. 2016;**6**:82244-82259

[54] Babuponnusami A, Muthukumar K. A review on Fenton and improvements to the Fenton process for wastewater treatment. Journal of Environmental Chemical Engineering. 2014;**2**:557-572

[55] Jia Z, Hao S, Liu Z. Synthesis of $BaSO_4$ nanoparticles with a membrane reactor: Parameter effects on membrane fouling. Journal of Membrane Science. 2017;**543**:277-281

[56] El-Ashtoukhy E-SZ, Amin NK, Abd El-Latif MM, Bassyouni DG, Hamad HA. New insights into the anodic oxidation and electrocoagulation using a self-gas stirred reactor: A comparative study for synthetic C.I reactive violet 2 wastewater. Journal of Cleaner Production. 2017;**167**:432-446

[57] Bassyouni DG, Hamad HA, El-Ashtoukhy E-SZ, Amin NK, Abd El-Latif MM. Comparative performance of anodic oxidation and electrocoagulation as clean processes for electrocatalytic degradation of diazo dye Acid Brown 14 in aqueous medium. Journal of Hazardous Materials. 2017;**335**:178-187

[58] Hamad H, Bassyouni D, El-Ashtoukhy E, Amin N, El-Latif MA. Electrocatalytic degradation and minimization of specific energy consumption of synthetic azo dye from wastewater by anodic oxidation process with an emphasis on enhancing economic efficiency and reaction mechanism. Ecotoxicology and Environmental Safety. 2018;**148**:501-512

[59] Hamad H, Bailón-García E, Morales-Torres S, Carrasco-Marín F, Pérez-Cadenas AF, Maldonado-Hódar FJ. Physicochemical properties of new cellulose-TiO_2 composites for the removal of water pollutants: Developing specific interactions and performances by cellulose functionalization. Journal of Environmental Chemical Engineering. 2018;**6**:5032-5041

[60] Hamad H, Bailón-García E, Maldonado-Hódar FJ, Pérez-Cadenas AF, Carrasco-Marín F, Morales-Torres S. Synthesis of TixOy nanocrystals in mild synthesis conditions for the degradation of pollutants under solar light. Applied Catalysis B: Environmental. 2019;**241**:385-392

[61] Hamad H, Abd El-latif M, Kashyout A, Sadik W, Feteha M. Optimizing the preparation parameters of mesoporous nanocrystalline titania and its photocatalytic activity in water: Physical properties and

growth mechanisms. Process Safety and Environmental Protection. 2015;**98**:390-398

[62] Srivastava A, Yadav T, Sharma S, Nayak A, Kumari AA, Mishra N. Journal of Biosciences and Medicines. 2015;**4**:69

[63] George A, Shah PA, Shrivastav PS. Natural biodegradable polymers based nano-formulations for drug delivery: A review. International Journal of Pharmaceutics. 2019;**561**:244-264

[64] Englert C, Brendel JC, Majdanski TC, Yildirim T, Schubert S, Gottschaldt M, et al. Pharmapolymers in the 21st century: Synthetic polymers in drug delivery applications. Progress in Polymer Science. 2018;**87**:107

[65] Damodaran VB, Bhatnagar D, Murthy NS. Manhattan, New Work City: Springer; 2016

[66] Parrott M, Dunn S. Design of Biomedical Polymers. Functional Biopolymers. 2018;**19**:1-48

[67] Yadav P, Yadav H, Shah VG, Shah G, Dhaka G. Biomedical Biopolymers, their Origin and Evolution in Biomedical Sciences: A Systematic Review. Journal of Clinical and Diagnostic Research. 2015;**9**:21-25

[68] Absar S, Khan M, Edwards K, Neumann J. Investigation of synthesis and processing of cellulose, cellulose acetate and poly(ethylene oxide) nanofibers incorporating anti-cancer/tumor drug cis-diammineplatinum (II) dichloride using electrospinning techniques. Journal of Polymer Engineering. 2015;**35**:867-878

[69] Gouda M, Hebeish A, Aljafari A. Synthesis and characterization of novel drug delivery system based on cellulose acetate electrospun nanofiber mats. Journal of Industrial Textiles. 2014;**43**:319-329

[70] Mehrabi F, Shamspur T, Mostafavi A, Saljooqi A, Fathirad F. Synthesis of cellulose acetate nanofibers and its application in the release of some drugs. Nanomedicine Research Journal. 2017;**2**:199-207

[71] Khoshnevisan K, Maleki H, Samadian H, Shahsavari S, Sarrafzadeh MH, Larijani B, et al. Cellulose acetate electrospun nanofibers for drug delivery systems: Applications and recent advances. Carbohydrate Polymers. 2018;**198**:131-141

[72] Ahmadian-Fard-Fini S, Ghanbari D, Amiri O, Salavati-Niasari M. Electro-spinning of cellulose acetate nanofibers/Fe/carbon dot as photoluminescence sensor for mercury (II) and lead (II) ions Carbohydrate Polymers. 2020;**229**:1-9

[73] Alghoraibi I, Alomari S. Handbook of Nanofibers. Cham, Switzerland: Springer; 2018. pp. 1-46

[74] Pant B, Park M, Park S-J. Drug Delivery Applications of Core-Sheath Nanofibers Prepared by Coaxial Electrospinning: A Review Pharmaceutics. 2019;**11**:1-29

[75] Shahriar SM, Mondal J, Hasan MN, Revuri V, Lee DY, Lee YK. Electrospinning Nanofibers for Therapeutics Delivery Nanomaterials. 2019;**9**:1-32

[76] Agrahari V, Agrahari V, Meng J, Mitra AK, Mitra AK, Cholkar K, et al., editors. In Emerging Nanotechnologies for Diagnostics, Drug Delivery and Medical Devices. first ed. Boston: Elsevier; 2017. pp. 189-215

[77] Puls J, Wilson SA, Holter D. Degradation of cellulose acetate-based

materials: A review. Journal of Polymers and the Environment. 2011;**19**:152-165

[78] Shalaby T, Hamad H, Ibrahim E, Mahmoud O, Al-Oufy A. Electrospun nanofibers hybrid composites membranes for highly efficient antibacterial activity. Ecotoxicology and Environmental Safety. 2018;**162**:354-364

[79] Jacob J, Haponiuk JT, Thomas S, Gopi S. Biopolymer based nanomaterials in drug delivery systems: A review Materials Today Chemistry. 2018;**9**:43-55

[80] Taepaiboon P, Rungsardthong U, Supaphol P. Vitamin-loaded electrospun cellulose acetate nanofiber mats as transdermal and dermal therapeutic agents of vitamin A acid and vitamin E. European Journal of Pharmaceutics and Biopharmaceutics. 2007;**67**:387-397

Chapter 4

Study on Pyrolysis Behaviors of Various Plant Fibers

Ke Zhang, Quanxing Zheng, Zhongya Guo, Lili Fu, Qi Zhang and Bing Wang

Abstract

Pyrolysis is an effective way to convert plant fibers into high-value-added chemicals and bioenergy. The pyrolysis behavior of plant fibers varies with their compositions. A high-performance anion-exchange chromatography integrated pulse amperometric method was established to detect the composition of arabinose, galactose, glucose, xylose, and mannose in plant fiber hydrolysate. The contents of cellulose, hemicellulose, and lignin in six plant fibers were calculated. Furthermore, the pyrolysis kinetic parameters of the plant fibers and their pyrolysis product distribution depending on chemical compositions were analyzed. The pyrolysis of flax fiber with high cellulose content (92.19%) tended to generate ketones, accounting for about 37.3% of the total product distribution, while coniferous and broadleaf fiber with high hemicellulose contents (13.23 and 15.07%, respectively) was more likely to generate aldehydes and hydrocarbons. Furthermore, the result of pyrolysis of a grass fiber demonstrated the interactions between its chemical components, which had been captured during pyrolysis from the perspective of pyrolysis product distribution that inhibits the pyrolysis to generate CO_2, and promoted the generation of furan, phenols, and toluene, to different degrees. The research results are expected to provide basic data and theoretical support for obtaining high-value-added chemicals and biomass energy through the pyrolysis of plant fibers.

Keywords: plant fiber, biomass chemical component, pyrolysis characteristic, pyrolysis products, pyrolysis behaviors

1. Introduction

Plant fiber is a kind of natural composite, which is mainly composed of cellulose, hemicellulose, and lignin [1]. The three polysaccharides are quite different in the composition of monosaccharides. Cellulose is a linear polymer composed of dehydrated glucose linked by β-1,4-glycosidic bonds. Hemicellulose is much more complex than cellulose. It is a heteropolysaccharide with some branches composed of pentose (xylose and arabinose), hexose (glucose, mannose, and galactose) and hexoic acid (4-O-methyl-D-glucuronic acid, D-glucuronic acid, and D-galacturonic acid). These functional groups can be assembled into various hemicellulose polysaccharides with different structures from linear to highly branched, such as β-1,4-D-xylan, arabinose xylan, mannan, dextran, galactose, and galactomannan. The detailed sugar compositions and chemical structures of these hemicellulosic polysaccharides vary according to plant species [1, 2]. Lignin is an aromatic polymer with highly branched chains, which is composed of phenylpropane derivative monomers (such as coumarin, coniferol, and sinapinol). The structure of lignin is complex and its molecular weight is large [1]. Due to the complex connection between cellulose, hemicellulose, and lignin in plant fibers, it is a challenge to complete the separation of the three components. At present, the "nitric acid-ethanol method," the "12% hydrochloric acid hydrolysis method," and the "72% sulfuric acid method" is used to measure the cellulose content, hemicellulose or pentosan content, and lignin content, respectively. However, these methods are cumbersome and the structural composition of hemicellulose could not be distinguished effectively. The National Renewable Energy Laboratory (NREL) of the United States has developed a method to systematically analyze the contents of cellulose, hemicellulose, and lignin in fibers by two-stage acid hydrolysis [3]. This method was improved by us to use high concentration sulfuric acid to convert plant fibers into oligosaccharides at low temperatures firstly, and then using dilute acid to further convert oligosaccharides into monosaccharides at high temperatures. The contents of cellulose and hemicellulose are determined based on analyzing monosaccharides by high-performance liquid chromatography. The contents of acid-insoluble lignin and acid-soluble lignin are determined by weighing the filter residue and detecting acid hydrolysate with UV. It is simple and easy to operate and is widely used by international research institutions [3, 4], and it is also economically viable for scaling up.

Moreover, pyrolysis of plant fibers is a multistep reaction process due to their complex multi-components. At present, calculation of pyrolysis kinetic parameters and analysis of pyrolysis product distribution from the perspective of cellulose, hemicellulose, lignin, and other component groups are effective ways to deeply understand plant fibers' pyrolysis behavior. The pyrolysis and combustion kinetics of various fibers are essential for the multipurpose utilization of biomass materials, because pyrolysis is an effective way to convert plant fibers into high-value-added chemicals and bioenergy [5–9]. Compared with model method, model-free kinetics, as Friedman method [10–14], Vyazovkin method [10, 13, 15], Ozawa method [16, 17], Kissinger-Akahira-Sunose (KAS) method [13, 15, 18], Flynn-Wall-Ozawa (FWO) method [13, 15, 18], and distributed activation energy model (DAEM) [11, 19–22], could reduce the errors introduced by the model fit. According to Friedman method [11, 13, 16], the pyrolysis and combustion processes of different plant fibers by using multiple heating rate programs were used to obtain reliable kinetics, in order to reveal their pyrolysis and combustion properties. In addition, the pyrolytic characteristics and tendentious pyrolysis reaction path of each polysaccharide (cellulose, hemicellulose, or lignin) could be demonstrated from the perspective of the pyrolytic product distribution of plant fibers with different chemical compositions. Interactions among these different chemical compositions during pyrolysis were also first noticed and speculated by analyzing the pyrolysis product of the actual whole and the sum of pyrolysis product by the accumulation of individual chemical composition. These pyrolysis properties investigated above would be useful references for the production of high-value-added products and biomass energy from plant fibers.

2. Analysis of plant fiber composition

High-performance anion-exchange chromatography with an integrated pulsed amperometric method (HPAEX-IPAM) is a newly developed method for sugar analysis in recent years. Zheng [23, 24] established HPAEX-IPAM for the

Samples	Cellulose	Hemicellulose	Lignin
Coniferous fiber	74.56 ± 0.23	13.23 ± 0.43	9.32 ± 1.61
Broadleaf fiber	75.80 ± 0.31	15.07 ± 0.55	5.55 ± 1.21
Bamboo fiber	74.05 ± 0.52	18.28 ± 0.46	5.86 ± 0.71
Flax fiber	92.19 ± 0.34	4.32 ± 0.16	2.33 ± 0.51
Grass fiber	71.90 ± 0.17	17.59 ± 0.55	4.28 ± 0.50
Cotton fiber	95.84 ± 0.97	0.23 ± 0.01	1.17 ± 0.01

Table 1.
Information of plant fiber compositions (%) [27].

determination of monosaccharides (arabinose, galactose, glucose, xylose, and mannose), glycuronate acid (galacturonic acid and glucuronic acid), and cellobiose in plant fiber hydrolysate. In order to hydrolyze cellulose, National Renewable Energy Laboratory (NREL) method optimized by the heating method is used to hydrolyze cellulose into monosaccharides. Then, monosaccharides are separated by an efficient anion exchange column in a strong alkaline medium, and then, the current generated by the oxidation reaction of hydroxyl groups in the sugar molecular structure on the gold electrode surface is detected [25, 26]. This method is simple in pretreatment, and the pulsed amperometric detector has high sensitivity, selectivity, accuracy, and reproducibility. Calculate the content of fiber components, according to the concentration of hydrolysis products of each fiber component. Zheng [23, 24] used this method to determine the acid hydrolysis monosaccharides of six common plant fibers (coniferous fiber, broadleaf fiber, bamboo fiber, flax fiber, grass fiber, and cotton fiber), and calculated the composition of cellulose and hemicellulose of each plant fiber, according to the corresponding hydrolysis monosaccharides mass fraction of cellulose and hemicellulose (as shown in **Table 1**). The lignin content shown in **Table 1** is the sum of acid-insoluble lignin and acid-soluble lignin, which are determined by weighing the filter residue and detecting acid hydrolysate with UV.

3. Pyrolysis behaviors of plant fibers

3.1 Pyrolysis characteristic of plant fibers

In general, the process of biomass pyrolysis includes four stages [5]. The first stage (<150°C) is dehydration, where the biomass releases external water by absorbing heat. This registers as a slight weight loss on the TG curve. The second stage (150–250°C) is evaporation and distillation of the volatile and semi-volatile components. At the same time, lignin begins to lose weight and releases some small molecules, and the crystalline regions of cellulose transform to amorphous partly. The major pyrolysis stage (250–500°C) shows the maximum weight loss peak on the DTG curve of the biomass. The last stage (>500°C) corresponds to the further slow decomposition of residues that generated from incomplete pyrolysis. **Figure 1** shows the TGA curve of coniferous fiber at a heating rate of 10°C/min and it presents the typical pyrolysis characteristics of biomass under inert atmosphere. The initial pyrolysis temperature T_s and terminal temperature T_h are obtained by the tangent method [28, 29], that are 339°C and 362.4°C, respectively.

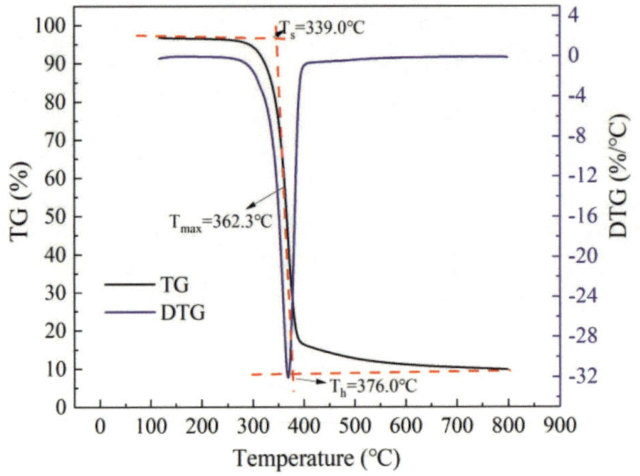

Figure 1.
TG/DTG curves of coniferous fiber sheet under N_2 atmosphere (heating rate β = 10°C/min) [27].

The pyrolysis behaviors of various plant fibers are quite distinct because of their different morphologies, chemical components, and pyrolysis conditions. **Table 2** demonstrates the pyrolysis characteristic parameters of various plant fibers. With an increase in heating rates, the values of these parameters T_s, T_h, and the peak temperature T_{max} show an increasing trend [27]. Zhao [30] investigated the pyrolysis rate elevated linearly with the increase in heating rates. The heating rate has an appreciable impact on the temperature difference for heat transfer and the temperature gradient between the measuring point and the sample, besides, that causes an additional endothermic amount to balance the thermal hysteresis. The distribution of cellulose, hemicellulose, and lignin demonstrates differences across species of plant fibers [31–33]. The pyrolysis temperature of the hemicellulose is the lowest, while that of lignin behaves the highest. As shown in **Table 2**, the maximum mass loss rate $-(dm/dt)_{max}$ of the fibers rises with the increase of the heating rate, which $-(dm/dt)_{max}$ of cotton fiber is the highest, and $-(dm/dt)_{max}$ of grass fiber is the lowest. When the heating rate is 15°C/min, the values of $-(dm/dt)_{max}$ are 35.2%/min and 24.1%/min, respectively. In addition, the pyrolysis index P [34, 35], which reflects the pyrolysis degree of fibers, enhances with the increase of heating rates.

3.2 Pyrolysis and combustion kinetics of plant fibers

As a typical biomass, the pyrolysis and combustion kinetics of various plant fibers are essential for the multipurpose utilization of biomass materials [6–9]. The TG/DTG data obtained at different heating rates (5, 10, and 15°C/min) were applied to get the relationship between $\ln\left[\beta\left(\frac{d\alpha}{dT}\right)_\alpha\right]$ and $\frac{1}{T_\alpha}$ for each plant fiber, at selected α values based on the Friedman method, as shown in **Figure 2**. The activation energy (E_a) corresponding to the α values could be obtained from the slope of the fitting curve, as shown in **Table 3** [27]. The value of α ranged from 0.05 to 0.95, and the step size was 0.05.

Due to the complicated pyrolysis processes of biomass, the apparent kinetics could be described by the appropriate models based on the global weight loss, while

Samples	β	T_s	T_h	T_{max}	$-(dm/dt)_{max}$	$\triangle T_{1/2}$	$P\times10^{-6}$
	/K·min^{-1}	/K	/K	/K	/(%·min^{-1})	/°C	/(%·min^{-1}·K^{-3})
Coniferous fiber	5	602.8	637.1	624.6	12.5	25.3	1.31
	10	612.0	649.3	635.4	22.9	27.9	2.11
	15	617.4	657.3	641.6	32.1	30.5	2.65
Broadleaf fiber	5	597.8	634.0	621.4	11.9	26.2	1.22
	10	608.9	648.1	634.1	22.2	29.1	1.98
	15	613.5	655.5	640.1	31.0	31.7	2.49
Bamboo fiber	5	595.2	633.3	619.2	11.0	27.7	1.08
	10	604.6	646.2	631.2	20.3	30.7	1.74
	15	609.3	652.4	636.5	29.1	32.6	2.30
Flax fiber	5	607.6	645.2	630.8	11.5	29.3	1.03
	10	618.3	658.3	643.2	21.8	31.7	1.73
	15	625.0	666.0	649.1	31.6	27.7	2.81
Grass fiber	5	591.7	637.8	619.8	9.0	33.0	0.74
	10	601.2	651.1	631.9	16.9	36.1	1.23
	15	607.4	659.8	639.7	24.1	39.0	1.59
Cotton fiber	5	610.0	643.4	629.8	13.2	27.1	1.27
	10	620.6	656.4	640.8	24.8	29.9	2.09
	15	627.1	664.0	648.1	35.2	31.4	2.76

β: heating rate; T_s: initial decomposition temperature; T_h: terminal decomposition temperature; T_{max}: peak temperature; $-(dm/dt)_{max}$: maximum mass loss rate; $\triangle T_{1/2}$: peak width at half-height; and P: pyrolysis index.

Table 2.
Pyrolysis characteristic parameters of fibers [27].

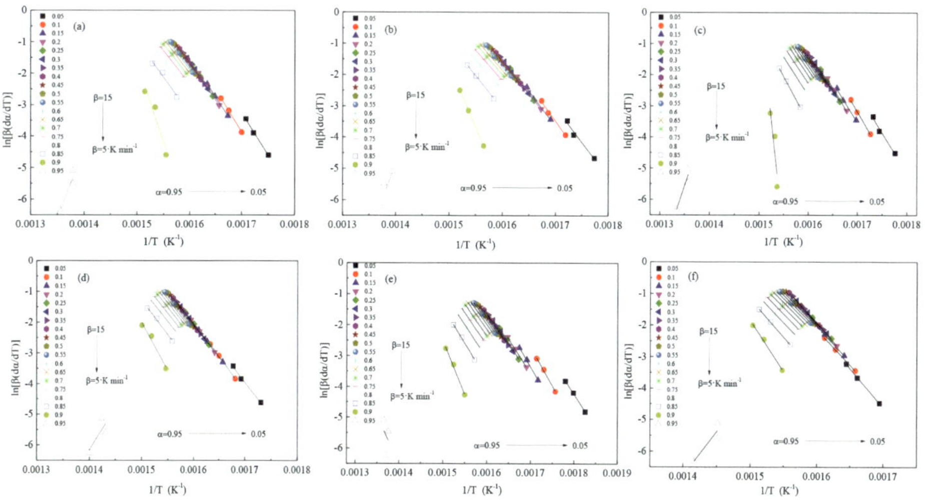

Figure 2.
Friedman results of pyrolysis of different fiber sheets under N_2 atmosphere (a) coniferous fiber, (b) broadleaf fiber, (c) bamboo fiber, (d) flax fiber, (e) grass fiber, and (f) cotton fiber [27].

α	Coniferous		Broadleaf		Bamboo		Flax		Grass		Cotton	
	E_α	R^2	E_α	R^2	E_α	R^2	E_α	R^2	E_α	R^2	E_α	R^2
0.05	224.16	0.9962	185.76	0.9769	214.39	0.9837	182.96	0.9952	183.16	0.9989	206.94	0.9994
0.1	228.85	0.9992	196.07	0.9927	227.60	0.9968	193.57	0.9997	209.72	0.9971	186.28	1.0000
0.15	211.04	1.0000	188.48	0.9910	216.08	0.9949	184.16	1.0000	203.67	0.9985	189.93	0.9998
0.2	210.38	0.9999	188.72	0.9954	218.02	0.9962	188.78	1.0000	184.92	0.9998	188.52	0.9996
0.25	203.58	0.9998	185.37	0.9963	213.00	0.9967	187.00	0.9999	186.89	0.9992	186.82	0.9995
0.3	197.55	0.9995	181.42	0.9972	205.12	0.9976	186.65	1.0000	182.30	0.9993	185.16	0.9998
0.35	193.79	0.9994	178.06	0.9980	201.10	0.9963	184.48	0.9999	176.78	0.9992	189.52	1.0000
0.4	190.80	0.9998	173.30	0.9980	198.15	0.9974	184.77	1.0000	173.92	0.9986	191.19	0.9990
0.45	188.00	0.9994	171.94	0.9983	193.86	0.9962	186.93	0.9999	170.33	0.9987	187.55	0.9994
0.5	186.02	0.9995	169.93	0.9986	193.62	0.9974	187.58	0.9996	168.08	0.9989	181.27	1.0000
0.55	184.37	0.9996	168.42	0.9977	192.50	0.9968	185.69	1.0000	168.19	0.9988	177.59	0.9999
0.6	179.99	0.9987	167.40	0.9982	191.93	0.9979	182.11	1.0000	167.02	0.9986	176.19	1.0000
0.65	177.07	0.9978	165.31	0.9989	191.79	0.9975	179.71	0.9996	167.47	0.9988	176.97	0.9999
0.7	174.73	0.9953	161.26	0.9995	192.96	0.9972	178.66	0.9991	170.44	0.9986	177.59	1.0000
0.75	175.14	0.9898	160.78	1.0000	196.99	0.9981	177.87	0.9987	173.45	0.9983	178.37	1.0000
0.8	177.50	0.9845	162.33	1.0000	205.95	0.9992	180.06	0.9975	179.63	0.9982	184.03	0.9999
0.85	194.01	0.9744	178.59	0.9994	245.74	1.0000	188.62	0.9935	194.08	0.9976	199.28	0.9998
0.9	422.16	0.9265	334.18	1.0000	1227.04	0.8234	261.73	0.9700	287.75	0.9931	267.37	0.9945
0.95	−378.69	0.8906	−396.67	0.5410	−400.47	0.9187	−227.41	0.6301	259.91	0.7083	−232.31	0.8507
E_α (α:0.05–0.85)	193.79		173.30		201.10		184.77		176.78		186.28	

Table 3.
The apparent activation energy E_α(kJ/mol) of different fibers pyrolysis under N_2 atmosphere obtained by Friedman method and correlation coefficient R^2 [27].

the pyrolysis mechanism of a specific component is still an intractable problem. Plant fiber is composed of complex multi-components and its pyrolysis is a multistep reaction process. As indicated in **Table 3**, the resulting apparent activation energies of plant fibers are various with the conversion rate. The average apparent activation energies of coniferous, broadleaf, bamboo, flax, grass, and cotton fibers are 193.79, 173.30, 201.10, 184.77, 176.78, and 186.28 kJ/mol, respectively, among the conversion range of 0.05–0.85. The average apparent activation energy of broadleaf fiber is the lowest and that of bamboo fiber is the highest.

Although biomass pyrolysis is generally carried out under an inert atmosphere, some research has been performed under an oxygen atmosphere [27]. The characteristic parameters, both of pyrolysis process of fibers and combustion of residual fixed carbon, as *Ts*, Th, and Tmax, demonstrate a high-temperature-shifting with the increase of the heating rate. The Tmax of all fibers behaves are lower than those under nitrogen condition, and the apparent activation energies show a similar variation trend, especially among the conversion between 0.05 and 0.65. Cotton fiber shows the highest Tmax, compared with grass fiber, which is the lowest. Improving the heating rate benefits the combustion performance of plant fibers and the combustion characteristic index (S) [27] increases accordingly. The result shows that the broadleaf fiber has the largest combustion characteristic index, while grass fiber has the smallest. The apparent activation energy of plant fiber pyrolysis in an oxygen atmosphere is lower than that in nitrogen, indicating that oxygen atmosphere can promote the pyrolysis reactions of plant fibers.

4. Pyrolytic product distribution analysis of plant fibers affected by chemical compositions and interactions among them

Pyrolysis behaviors of coniferous, broadleaf, bamboo, flax, grass, and cotton fibers have been proven to be relevant with the biomass types and chemical compositions [5], showing different pyrolysis kinetic parameters because of varieties of cellulose and hemicellulose compositions in diverse plant fibers [27]. The distribution of pyrolysis products would also be largely restricted by the different chemical compositions of these fibers. The pyrolysis products and their contents of flax fiber, coniferous fiber, and broadleaf fiber at 350°C are shown in **Table 4** [36]. The main pyrolysis products of several plant fibers included alcohol, aldehyde, ketone (furanone), acid, ester, hydrocarbon, anhydrosugar, and CO_2. Flax fiber contained a high proportion of cellulose (92.35%), hence, a high proportion of glucose molecular units and its ketone yield in the pyrolysis product is the highest (37.3%). The glucose ring could be easily broken at C_1-O and C_2-C_3 confirmed by Piskorz et al. [37] based on bond energy analysis, forming fragments of two carbon atoms (C_1/C_2) and four carbon atoms (C_3–C_6), of which two-carbon fragments could form hydroxyacetaldehyde, while four-carbon fragments would generate hydroxyacetone and other products after thermal decomposition process. In addition, 1,3-dihydroxy-2-propanone mainly originates from the breaking of the cellulose chain during pyrolysis [38]. Compared with flax fiber, the cellulose contents of coniferous fiber and broadleaf fiber are significantly reduced, the hemicellulose contents are significantly increased, and their ketone yields after pyrolysis show significantly reduced, while the yields of aldehydes and hydrocarbons are enhanced. It could be related to the fact that the pyrolysis of five-carbon sugars (xylose and arabinose) tended to generate furfural, and the pyrolysis of six-carbon sugars (glucose, mannose, and galactose) tended to generate

Pyrolysis products percent (%)	Aldehydes	Ketones	Hydrocarbons	Esters	Alcohols	Anhydrosugars	Acids	CO_2
Flax fiber	17.9	37.3	11.1	7.0	11.7	0.5	2.4	9.7
Coniferous fiber	24.3	13.6	15.6	14.1	12.9	0.5	6.2	7.5
Broadleaf fiber	23.8	20.6	15.1	7.9	9.5	0.4	4.0	10.2

Table 4.
Pyrolysis product distribution of flax fiber, coniferous fiber, and broadleaf fiber at 350°C [36].

5-hydroxymethylfurfural [3, 39–41]. 5-hydroxymethylfurfural is mainly obtained through an acetal reaction between C-2 and C-5, furfural can be obtained through the hydroxymethyl elimination reaction of 5-hydroxymethylfurfural [38], or through the cyclization/dehydration reaction of xylose [39].

Hemicellulose and lignin of lobular seal could be obtained through step-by-step extraction by Xuefei Cao [42], the main components of water-soluble hemicellulose are β-D-glucan and a small amount of pectin. The main component of alcohol-soluble hemicellulose is poly arabinogalactose. The main pyrolysis range of cellulose and hemicellulose spans 200–400°C, and their pyrolysis products are mainly carbonyl compounds. The main pyrolysis range of lignin is 300–700°C, and its pyrolysis products are mainly aromatic compounds. Both glycosidic bonds of cellulose and hemicellulose and the C-C bonds on the sugar ring have low dissociation energy, which is easy to break during pyrolysis to produce low carbon and oxygen-containing small molecule products. The bond dissociation energies of the methyl group on methoxy group of lignin monomer, C_α-C_β connection bond on side chains, as well as β-O-4′ ether bond and C_α-C_β in lignin dimer connection are small, so the lignin pyrolysis is easy to produce phenolic products with short side chains.

Therefore, the distribution of fiber pyrolytic products is closely related to their differences in cellulose/hemicellulose composition, and maybe the interactions between them as well. The production and utilization of reconstituted tobacco leave based on component reconstruction involved the separation and reorganization/reassembling of water-impregnated extracts, ethanol-impregnated extracts, etc. of tobacco biomass. We take a grass fiber as an experimental example, its water-soluble components, ethanol-soluble components, and residual solid-phase components could be obtained through a step-by-step extraction process. The actual pyrolysis product distribution of the grass fiber and the weighted pyrolysis product distribution based on its component distribution are investigated, and the comparison results confirm that there are interactions between the chemical components of the plant fiber during pyrolysis [43], in which the generation of CO_2 is inhibited by the interactions, and the generation of furan, phenol, and toluene was promoted by the interactions between components to varying degrees.

5. Conclusions

In this chapter, a method for the determination of arabinose, galactose, glucose, xylose, mannose, and other monosaccharides in plant fiber hydrolysate is established based on high-performance anion exchange chromatography integrated pulse amperometric technique. The acid hydrolysis monosaccharides in coniferous fiber, broadleaf fiber, bamboo fiber, flax fiber, grass fiber, and cotton fiber could be determined by this method, and cellulose, hemicellulose, and lignin contents of each plant fiber were further calculated. Plant fibers are complex polymers, and their pyrolysis process is very complex. Under inert atmosphere, the average apparent activation energies of coniferous fiber, broadleaf fiber, bamboo fiber, flax fiber, grass fiber, and cotton fiber with different chemical compositions in the conversion rate of 0.05–0.85 are 193.79, 173.30, 201.10, 184.77, 176.78 and 186.28 kJ/mol, respectively. The apparent activation energies of plant fibers pyrolysis in the oxygen atmosphere are lower than those in the nitrogen atmosphere. The oxygen atmosphere can promote the pyrolysis of plant fibers. The pyrolysis of flax fiber with high cellulose content tends to generate ketones, while coniferous fiber and broadleaf fiber with high hemicellulose contents are more likely to generate aldehydes and hydrocarbons. Furthermore, taking a grass fiber as the experimental object, interactions between its chemical components have been captured during pyrolysis from the perspective of pyrolysis product distribution, which inhibits the pyrolysis to generate CO_2, and promote the generation of furan, phenols, toluene, etc. to different degrees.

Conflict of interest

The authors declare no conflict of interest.

Acronyms and abbreviations

NREL	National Renewable Energy Laboratory
E_α	Average apparent activation energy
KAS	Kissinger-Akahira-Sunose method
FWO	Flynn-Wall-Ozawa method
DAEM	Distributed activation energy model
HPAEX-IPAM	High-performance anion-exchange chromatography with integrated pulsed amperometric method

Author details

Ke Zhang[1], Quanxing Zheng[2], Zhongya Guo[3], Lili Fu[1], Qi Zhang[1] and Bing Wang[1]*

1 Zhengzhou Tobacco Research Institute of CNTC, Zhengzhou, China

2 China Tobacco Fujian Industrial Co., Ltd., Xiamen, China

3 China Tobacco Guangdong Industrial Co. Ltd, Guangzhou, China

*Address all correspondence to: wangbing6696@126.com

References

[1] Shi SL, He FW. Analysis and Detection of Pulp and Papermaking. Beijing: China Light Industry Press; 2003

[2] Wang S, Ru B, Lin H, et al. Pyrolysis behaviors of four O-acetyl-preserved hemicelluloses isolated from hardwoods and softwoods. Fuel. 2015;**150**:243-251. DOI: 10.1016/j.fuel.2015.02.045

[3] Jiang L, Wu N, Zheng A, et al. Comprehensive utilization of hemicellulose and cellulose to release fermentable sugars from corncobs via acid hydrolysis and fast pyrolysis. ACS Sustainable Chemistry Engineering. 2017;**5**:5208. DOI: 10.1021/acssuschemeng.7b00561

[4] Moxley G, Zhang Y. More accurate determination of acid-labile carbohydrates in lignocellulose by modified quantitative saccharification. Energy & Fuels. 2007;**21**:3684. DOI: 10.1021/ef7003893

[5] Wang SR, Luo ZY. Pyroloysis of Biomass. Beijing: Science Press; 2016

[6] George WH, Juben NC, Christopher JB, et al. Production of liquid alkanes by aqueous-phase processing of biomass-derived carbohydrates. Science. 2005;**308**(5727):1446-1450. DOI: 10.1126/science.1111166

[7] George WH, Sara I, Avelino C. Synthesis of transportation fuels from biomass: Chemistry, catalysts, and engineering. Chemical Reviews. 2006;**106**:4044-4098. DOI: 10.1021/cr068360d

[8] Juben NC, George WH, James AD. Liquid-phase catalytic processing of biomass-derived oxygenated hydrocarbons to fuels and chemicals. Angwwandte Chemie-International Edition. 2007;**46**:7164-7183. DOI: 10.1002/anie.200604274

[9] David MA, Stephanie GW, James AD. Bimetallic catalysts for upgrading of biomass to fuels and chemicals. Chemical Society Reviews. 2012;**41**:8075-8098. DOI: 10.1039/C2CS35188A

[10] Vyazovkin S. Modification of the integral isoconversional method to account for variation in activation energy. Journal of Computational Chemistry. 2001;**22**(2):178-183. DOI: 10.1002/1096-987X(20010130)22:2<178::AID-JCC5>3.0.CO;2-%23

[11] Hu M, Chen Z, Wang S, et al. Thermogravimetric kinetics of lignocellulosic biomass slow pyrolysis using distributed activation energy model, Fraser-Suzuki deconvolution, and iso-conversional method. Energy Conversion and Management. 2016;**118**:1-11. DOI: 10.1016/j.enconman.2016.03.058

[12] Sbirrazzuoli N. Determination of pre-exponential factors and of the mathematical functions $f(\alpha)$ or $G(\alpha)$ that describe the reaction mechanism in a model-free way. Thermochimica Acta. 2013;**564**:59-69. DOI: 10.1016/j.tca.2013.04.015

[13] Mishra G, Bhaskar T. Non isothermal model free kinetics for pyrolysis of rice straw. Bioresource Technology. 2014;**169**:614-621. DOI: 10.1016/j.biortech.2014.07.045

[14] Zhao ZL, Li HB, Yu C, Chen Y. Study on the kinetic characteristics of bagasse pyrolysis and combustion. Journal of Fuel Chemistry and Technology. 2005;**33**(3):314-319. DOI: 10.1016/j.molcatb.2005.02.001

[15] Collazzo GC, Broetto CC, Perondi D, et al. A detailed non-isothermal kinetic study of elephant grass pyrolysis from different models. Applied Thermal Engineering. 2017;**110**:1200-1211. DOI: 10.1016/j.applthermaleng.2016.09.012

[16] Ma Z, Wang J, Yang Y, et al. Comparison of the thermal degradation behaviors and kinetics of palm oil waste under nitrogen and air atmosphere in TGA-FTIR with a complementary use of model-free and model-fitting approaches. Journal of Analytical and Applied Pyrolysis. 2018;**1**(34):12-24. DOI: 10.1016/j.jaap.2018.04.002

[17] Duan L, Chen J, Jiang Y, et al. Experimental and kinetic study of thermal decomposition behaviour of phytoremediation derived Pteris vittata. Journal of Thermal Analysis & Calorimetry. 2016;**128**(2):1207-1216. DOI: 10.1007/s10973-016-6032-3

[18] Damartzis T, Vamvuka D, Sfakiotakis S, et al. Thermal degradation studies and kinetic modeling of cardoon (Cynara cardunculus) pyrolysis using thermogravimetric analysis (TGA). Bioresource Technology. 2011;**102**(10):6230-6238. DOI: 10.1016/j.biortech.2011.02.060

[19] Guo G, Iiu LC, Wang Y, et al. Comparative investigation on thermal degradation of flue-cured tobacco with different particle sizes by a macro-thermogravimetric analyzer and their apparent kinetics based on distributed activation energy model. Journal of Thermal Analysis & Calorimetry. 2019;**138**(5):3375-3388. DOI: 10.1007/s10973-019-08215-7

[20] Guo G, Zhang K, Liu C, et al. Comparative investigation on thermal decomposition of powdered and pelletized biomasses: Thermal conversion characteristics and apparent kinetics. Bioresource Technology. 2020;**301**:1-9. DOI: 10.1016/j.biortech.2020.122732

[21] Chen D, Zheng Y, Zhu X. In-depth investigation on the pyrolysis kinetics of raw biomass. Part I: Kinetic analysis for the drying and devolatilization stages. Bioresource Technology. 2013;**131**:40-46. DOI: 10.1016/j.biortech.2012.12.136

[22] Cai J, Wu W, Liu R, et al. A distributed activation energy model for the pyrolysis of lignocellulosic biomass. Green Chemistry. 2013;**15**:1331-1340. DOI: 10.1039/C3GC36958G

[23] Zheng QX, Zhang JP, Li QL, et al. Application of ion chromatography-integrated pulsed amperometric method in the analysis of monosaccharide composition of pulp fiber. China Pulp & Paper. 2020;**39**(7):37-43

[24] Zheng QX, Liu XC, Zhang GQ, et al. Quantifying cellulose, hemicellulose and lignin contents in common cigarette paper pulps. Tobacco Science Technology. 2020;**53**(9):40-46

[25] Chai Y, Wang H, Liu YM, et al. Methodology study of monosaccharide composition by HPAEC-PAD and its application. Pharmaceutical Biotechnology. 2016;**23**(4):332

[26] Le SF, Wang W, Wang YX, et al. Detection of seven monosaccharides in aloe polysaccharide using ion chromatography with pulsed amperometric detector. Chinese Journal of Chromatography. 2019;**37**(3):319

[27] Zheng QX, Liu XC, Wu TW, et al. Study on pyrolysis and combustion characteristics of different plant fibers. Journal of Fuel Chemistry and Technology. 2022;**50**(6):747-756

[28] Ma BG, Li XG, Xu L, et al. Investigation on catalyzed combustion

of high ash coal by thermogravimetric analysis. Thermochimica Acta. 2006;**445**(1):19-22

[29] Qin GX, Li B, Lu DF, et al. Combustion property and mechanism of tobacco biomass. Tobacco Science Technology. 2015;**48**(1):76-81

[30] Zhao S, Tan WY, Wang SY, et al. Influence of temperature programmed rate on pyrolysis rate of the stalk of soybeans and white birch. Journal of Northeast Forest University. 2003;**31**(2):75-77

[31] Zhou S, Xu YB, Wang CH, et al. A comparative study of the combustion behavior and mechanism of cellulose, pectin and starch. Acta Tabacaria Sinica. 2011;**17**(5):1-9. DOI: 10.1016/S1872-5805(11)60064-4

[32] Lü DZ, Hong Y, Wang QB, et al. Effect of cellulose and lignin content on pyrolysis and gasification characteristics for several types of biomass. American Journal of Physiology Regulatory Integrative & Comparative Physiology. 2009;**297**(1):R6-R16. DOI: 10.1152/ajpregu.00097.2009

[33] Zhao S, Xiao J, Shen LH, et al. Experimental study of biomass rapid pyrolysis based on three components. Acta Energiae Solaris Sinica. 2011;**32**(5):7710-7717. DOI: 10.1007/s12182-011-0123-3

[34] Gong DH, Xu C, Gu HY. Analysis on pyrolysis characteristics of tobacco stem. Journal of Guizhou Universitry (Natural Sciences). 2011;**28**(4):33-36

[35] Yu J, Zhang MC, Shen Y, et al. Thermogravimetric analysis of pyrolysis characteristics of biomass. Journal of Shanghai Jiaotong Universtry. 2002;**36**(10):1475-1478

[36] Zheng QX, Li QL, Zhang K, et al. Pyrolysis process of plant fibers. Chinese Journal of Applied Chemistry. 2022;**39**(7):1073-1082

[37] Piskorz J, Radlein D, Donald S. On the mechanism of the rapid pyrolysis of cellulose. Journal of Analytical and Applied Pyrolysis. 1986;**2**(9):121-137. DOI: 10.1016/0165-2370(86)85003-3

[38] Lu Q, Yang XC, Dong CQ, et al. Influence of pyrolysis temperature and time on the cellulose fast pyrolysis products: Analytical Py-GC/MS study. Journal of Analytical & Applied Pyrolysis. 2011;**2**(92):430-438. DOI: 10.1016/j.jaap.2011.08.006

[39] Zhou XW, Li WJ, Mabon R, et al. A mechanistic model of fast pyrolysis of hemicellulose. Energy & Environmental Science. 2018;**11**(5):1240-1260. DOI: 10.1039/C7EE03208K

[40] Zhou XW, Li WJ, Mabon R, et al. A critical review on hemicellulose pyrolysis. Energy Technology. 2017;**5**(1):52-79. DOI: 10.1002/ente.201600327

[41] Lin YC, Cho J, Geoffrey A, et al. Kinetics and mechanism of cellulose pyrolysis. Journal of Physical Chemistry C. 2009;**113**(46):20097-20107. DOI: 10.1021/jp906702p

[42] Cao XF. Separation, Modification and Thermochemical Conversion of the Components of Lignocellulose. PhD dissertation. South China University Technology, Chinese National Knowledge Infrastructure dissertation database; 2015. Available from: https://kns.cnki.net/KCMS/detail/detail.aspx?dbname=CDFDLAST2016&filename=1015986942.nh

[43] Guo Z, Zhang K, Zhang Q, et al. Biomass fractionation and its effects on pyrolysis chemistry. Journal of Analytical and Applied Pyrolysis. 2022;**167**:105650. DOI: 10.2139/ssrn.4069235

Section 2

Industrial Applications

Chapter 5

Advances in Cellulose-Based Packaging Films for Food Products

Eda Ceren Kaya and Umut Yucel

Abstract

Cellulose and its derivatives can be used to manufacture packaging film materials with versatile properties as alternatives to petroleum-based films. This chapter covers the recent trends and advancements in cellulose-based films for food materials. The chapter starts with the introduction of traditional and novel cellulose structures relevant to film-making properties including cellulose fibers, filaments, nano-fibrils, crystalline cellulose, and other traditional cellulose derivatives. The relevant cross-linking methods, such as esterification, etherification, oxidation, and carboxylation will be described in the production of materials such as methylcellulose, cellulose acetate, rayon fabric, carboxymethyl cellulose, cellulose ether, etc. The chapter will relate the properties of the films (e.g., crystallinity, mechanical, optical, barrier, and solubility properties) to the chemical characteristics of the cellulose materials. The chapter will also cover the interactions of cellulose with polymeric composites such as protein, polysaccharides, and other nanoparticles ingredients with a focus on emerging technologies.

Keywords: packaging films, cellulose structures, cellulose derivatives, physical modifications, chemical modifications, cellulose crosslinking, cellulose interactions

1. Introduction

The packaging film materials are significant contributors to safety, quality, and shelf life of food materials. They are as primary, secondary and tertiary depending on the level of contact with the product. Primary packaging films that are the focus of this chapter directly contact with products. Packaging films are made of a polymer matrix to create cohesive structure providing strength, often with a plasticizer to reduce rigidity and brittleness. They provide protection against physical and biological damage to food, and extent shelf life by limiting the diffusion of moisture, gases, volatile compounds.

Traditional packaging films are manufactured from petroleum-based polymers with tunable properties, but well-known environmental drawbacks. For example, the concentration of microplastics fragmented from synthetic films has been increasing to critical levels and is predicted to further increase in the next decades. The microplastic contamination in aquatic environment and soils considered as dominant environmental plastic pollutants to result in serious health concern for humans and animals [1]. The adoption of biodegradable films, which are known as materials that

IntechOpen

can naturally decompose into water, carbon dioxide, methane, or biomass in three to six months, can serve as a viable solution to this problem [2]. These materials can be obtained from biomass, synthesized from bio-derived monomers or produced from microorganisms [3]. Moreover, these materials can be produced from renewable resources (i.e., cellulosic materials and fibers recovered from process waste-streams) supporting sustainable agriculture practices [4]. There is, therefore, a large need for designing biodegradable, safe, and high-performance packaging materials using biopolymer sources, such as cellulose.

Biopolymers are natural macromolecules obtained from living organisms. They can be used to produce packaging films as alternatives to petroleum-based materials. The literature has plethora of examples for such films produced from polysaccharides, proteins, lipids, and their combinations [5]. Some of the biggest challenges for the adoption of biopolymers films are related to their limited film forming abilities, and lower mechanical and gas barrier properties as compared to traditional synthetic materials. The high production cost is another limitation related to the extraction, purification, conversion, and recovery steps involved.

Cellulose is the most abundant natural biopolymer found in cell walls of all plants as well as algae and fungi. It can serve as an ideal material for film manufacturing related to its versatility, chemical stability, non-toxicity, and affordability. The versatility of cellulose is also related to its various derivatives with distinct chemical structures and physicochemical properties used in a wide array of industrial segments as covered in this chapter. For example, cellulose acetate is one of the well-known cellulose derivatives and its commonly used to produce strong and resistant films in addition to distinct applications as frame materials in personal care items and other rigid tools (glasses, combs, screwdriver handles, etc.), imaging and photography films, filters, reservoirs for liquids, synthetic fiber etc. Therefore, it is safe to claim that cellulose and its derivatives are popular in packing applications. This chapter starts with the structures and sources of cellulose relevant to film-forming characteristics; then, continues with chemical and physical approaches to tailor cellulose derivatives. The chapter also highlights various polymeric interactions within cellulose fibrillar network. The chapter adds the use of cellulose in various food film applications with advances in manufacturing cellulose-based packaging films and emerging technologies. To illustrate prior knowledge, four recent review papers were included throughout the chapter [3, 6–8].

2. Cellulose structure and sources

2.1 Structure

The simplest and well-known definition of cellulose considers the polymer as D-glucopyranose subunits linked by β-1,4-glycosidic bond $(C_6H_{10}O_5)_n$. The degree of polymerization (n) of native cellulose expresses the number of glucose units that range between 1000 and 30,000 units depending on the source. For example, native wood cellulose of 10,000 glucose units is smaller than cotton cellulose of 15,000 units. Cellulose structure is typically in a linear form, where adjacent glucose units rotate by 180° to create subunits of a repeating cellobiose disaccharide. Larger organization of these subunits forms fibrous structures, commonly referred to as microfibrils or lignocellulosic fibers. Microfibril subunits are organized to generate cellulose fiber structure. The micro-scale interactions of cellulose and associated bonds have significant effect on its characteristics and functionality. For example, hydrogen bonds between three hydroxyl groups of the anhydroglucose unit and D-glucopyranose oxygen atom provide rigidity and generate the three-dimensional arrangement of cellulose in their polymer network. Besides, strong hydrogen bonding of the hydroxyl groups is responsible for the reduced solubility, crystallinity, and reactivity of the polymer.

The microfibril regions are composed of crystalline and amorphous regions. The crystalline structure formed by hydrogen bonding between adjacent microfibrils. Crystalline units are found in two different phases, namely, α-cellulose and β-cellulose. The predominance of crystalline phase is determined by its source (i.e., plant- and animal-based celluloses are predominantly in the β-cellulose form, and the bacteria-based cellulose is in α-form). The crystalline cellulose appears as rod-shaped particles, which can be extracted using enzymatic, chemical, or mechanical methods. The amorphous cellulose lack a certain shape and order, and formed by randomly ordered microfibrils surrounded by hemicelluloses and. In the amorphous region, the cellulose microfibrils are more available for hydrogen bond and subsequent interactions with water, protein or enzymes. Overall, the composition of crystalline and amorphous regions determine the physicochemical properties (e.g., rigidity, strength, extractability, etc.) of cellulose. For example, crystalline cellulose may display deuteration whereas amorphous cellulose swells in hydrophilic solvent and can penetrate inside the amorphous matrix by disrupting the intermolecular hydrogen bonds [9]. The high crystallinity can increase the rigidity and reduce the elasticity of cellulose-based films. Furthermore, these structures can be modified to obtain derivatives with desired functionalities as explained in Section 3.

2.2 Sources

Plants are the most common source of cellulose and cellulose derivatives. A schematic diagram of cellulose sources and modification methods is given **Figure 1**. The literature and industry have various examples for the use of various cellulose sources in packaging applications. Following extraction, cellulose is typically purified to eliminate lignin and hemicellulose for the manufacturing of cellulose derivatives. Among plant sources, cotton fibers have the advantage of having relatively low non-cellulosic components with around 90 wt% purity with long chains of crystalline regions (60%). Cellulose fibers from cotton displays twisted morphology with an internal structure of cell wall layers. The strength of cotton fiber increases with moisture. Therefore, cotton-based cellulose are suitable for film applications, where mechanical integrity is required. The second most abundant source of cellulose is wood. It displays not only good mechanical properties but also potential antibacterial characteristics due to the presence of natural phytochemicals [6]. Cellulose extracted from wood includes lignin and other polysaccharide compounds with ca. 50 wt% cellulose content. The presence of hemicellulose aids extraction of wood fibers easily during the pulping process.

Bacteria-based cellulose or cellobiose is produced by Gram (−) aerobic bacteria such as *Acetobacter xylinum*, *Gluconacetobacter xylinus*, *Komagataeibacter xylinus*, *Agrobacterium*, *Achromobacter*, *Pseudomonas*, and *Rhizobium*, and Gram (+) bacteria such as *Sarcina spp*. The bacteria-based cellulose is typically of high purity with ribbon-like shaped microfibril structures. Isolation and purification of bacterial cellulose do not require additional chemical treatment unlike plant-based cellulose. The bacteria-based cellulose microfibrils may agglomerate into a gel-like network (i.e.,

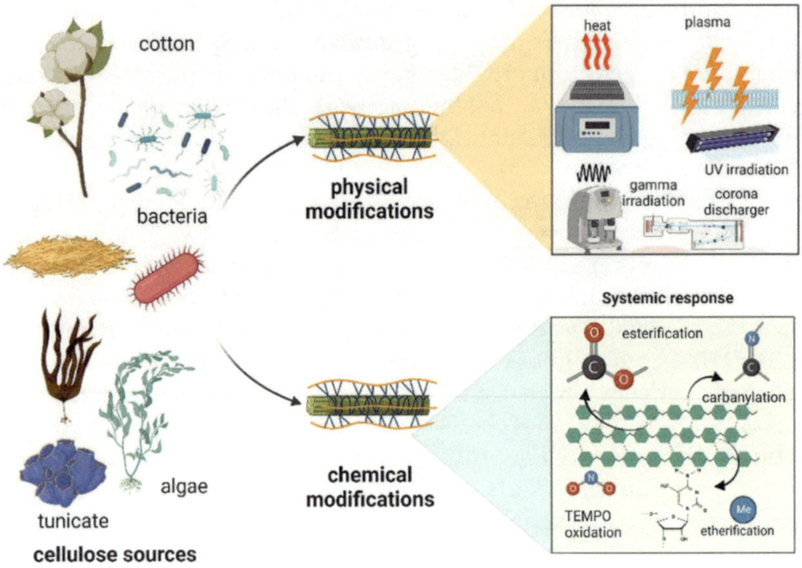

Figure 1.
A diagram of common cellulose sources and modification techniques.

microfibril gels) via interfibrillar hydrogen bonds and Van der Waals interactions, ideal for obtaining natural cellulose packaging films [10]. Another related structure is bacterial nanofibrils, which are thinner than microfibrils with higher surface area/volume ratio. This structure is desirable for film manufacturing since it can impart higher elasticity and strength. The microfibril structure can be controlled by altering the bacterial growth conditions. For example, using different carbon sources (e.g., replacing glucose with fructose, galactose, mannitol or xylose) can modify the surface area, porosity and crystallinity index of cellulose fibrils, and subsequently enhance resistance of their cellulose films against the oxygen and water vapor permeabilities [11]. Therefore, bacterial cellulose serves as a suitable source of cellulose with tunable characteristics for film applications.

Algae are known as renewable cellulose sources due to their ability to get nutrients from waste streams. Several species of green, gray, red, yellow, or brown algae, such as *Caldophora*, *Micrasterias denticulata*, *Gelidium elegans*, *Micrasterias rotate*, *Valonia*, *Boergesenia* and *Posidonia Oceanica*, were used to obtain cellulose microfibrils from their cell walls. Red algae contain the largest amount of cellulose. In comparison, green algae cellulose shows higher crystallinity (above 70%) than others, and therefore lower moisture adsorption capacity and limited the swelling. The purity of algae cellulose is typically lower than bacterial and plant cellulose. This is related to the presence of cellular proteins in the microfibril isolates. Algae based cellulose was used for biopolymer film applications, but they exhibited poor water vapor permeability and low mechanical strength [12]. Moreover, crystalline cellulose extracted from seaweed species of *Alaria esculenta*, *Saccharina latissima* have higher pure cellulose content as compared to cellulose isolated from *Ascophyllum nodosum* with heterogeneous content with minerals and proteins. Impurities reduces the presence of free hydroxyl group to limit their applications. The researchers showed that films produced from high purity cellulose exhibit better visual appearance and less moisture permeability [13].

Tunicates are invertebrate marine animals that can serve as a unique animal source of cellulose. Tunicate has three subclasses: *Ascidiacea*, *Thaliacea* and *Appendicularia.* Tunicate epidermal cells contain many enzyme complexes in epidermis membrane, which are responsible for cellulose production. The tunicate cellulose can be isolated nanofibril form, which are bundled in the outer tissues of tunicates. Tunicate cellulose exhibits higher crystallinity (ca. 95%) and surface area compared to other cellulose types. This provides excellent mechanical and thermal properties for cellulose-based films [14].

3. Modification of cellulose and cellulose derivatives

Cellulose is often modified via physical and chemical techniques to increase their functionality responsive to the film application and industry. Physical methods refer to approaches, which do not largely rely on the use of chemical compounds to modify cellulose fibers. The physical modification section covers the use of techniques that utilize mechanical, thermal, electrical, and other high-energetic (e.g., gamma irradiation, UV light) processes. The chemical modifications section covers common techniques to obtain traditional and novel cellulose derivatives including hydrolysis, esterification, acetylation, etherification, silylation, carbamylation, TEMPO oxidation. A summary of the modification methods for obtaining cellulose derivatives is illustrated in **Figure 1**.

3.1 Physical modifications

Thermal treatment can be used as a pre-treatment or post-treatment to remove undesired substances, enhance the oxygen-carbon ratio, and improve crystallinity of cellulose fibers. These modifications can improve the gas barrier properties and increase the acidity of the fiber surfaces used to produce films. For example,

heating (up to 145°C) tempo-oxidized cellulose nanofiber films was shown to significantly reduce the water retention and oxygen permeability as compared to original films prepared without post-thermal treatment [15]. Moreover, heating can help in inducing the closure of pores between cellulose nanofibers. This can be explained with the reorientation of amorphous and paracrystalline regions into the crystalline regions [16].

Plasma surface treatment can be used to enhance the hydrophilicity to cellulose fibers, and subsequently to provide functional properties to their films, such as better oxygen and air permeability. This technique utilizes electric discharge to generate vacuum-based ionized gas at ambient temperatures. The ionized gas can be generated via low pressure, low temperature, atmospheric glow discharge, atmospheric pressure plasma jet technique. The ionized gas includes a mixture of free radicals, heavy particles, and electrons, which promote the modification cellulose fibers and fiber aggregates as in films. The modifications can be in the form of defibrillation, remove low molecular weight impurities from nanocellulose fibers, and enhance microcrystalline cellulose [17]. Plasma treatment can increase the hydrophilicity of the fibers by generating polar functional groups, which provides good adhesion of cellulose coating layers. Plasma treatment was also shown to increase the crosslinking and entanglements of cellulose nanofibrils, increase the extent of impermeable crystals, which hinders the molecular diffusion of oxygen to increase the oxygen barrier properties of microfibrillar cellulose films [18].

The corona discharger process is similar to plasma surface treatment and involves the use of a bundle of charged particles, such as ions and electrons, accelerated through an electric field. The ionization process involves a shade of plasma that occurs at the tip of the particle generator that is named corona. This method has several advantages such as high efficiency, continuous processing and being environmentally friendly [19]. Cellulose fibers exposed to electromagnetic field with high voltage become excited and available for further surface modification via oxygenation. Electric discharge induces oxygenation of ozone into carbonyl groups that brings about crosslinking between fibers and the media to improve the mechanical and thermal properties of cellulose fiber films. The corona treatment process was shown to improve the interfacial adhesion between cellulose fiber and polypropylene in laminated film applications in addition to improved film strength and rigidity [19]. In parallel, corona treatment was also shown to enhance the strength and thermal resistance of cellulose films [20].

Gamma irradiation is a non-thermal and environmentally friendly process that modifies the chemical structure of cellulose fibers via high energetic gamma rays. High energy gamma irradiation results in decomposition and crosslinking, thus strengthening the structure of cellulose fibers [21]. Gamma rays typically are generated by using Cobalt-60 or Cesium-137 sources at specific irradiation facilities, which constitute a major limitation for its availability and wide-spread use. The irradiation process has several advantages over traditional treatment methods, such as no need for catalysts, less atmospheric pollution, minimum time for the process and continuous operation. Irradiation may cause initial structure disruption and irreversible bond cleavage, thus fragmentation of the polymers into nanofibrils. The high energetic of the gamma irradiation can result in crosslinking of cellulose nanofibrils into fiber networks and gels without the use of chemical reagents.

The mechanism involves the formation of macro-cellulosic radicals by the removal of hydrogen-hydroxyl and disturbance of carbon-carbon bonds through high energy gamma rays. These radicals are very reactive to facilitate cross-linking between cellulose nanocrystals and result in strong cellulose nanocrystal films with desirable oxygen and water vapor resistance [22].

UV radiation treatments is the application of electromagnetic radiation at wavelengths from 100 nm to 400 nm. This method uses ultraviolet laser light source to treat cellulose with enough energy to modify the molecular bonds on the surface layer. Briefly, UV radiation increases the formation of carbonyl groups on cellulose fiber surface and increases the surface polarity. In addition to this, the strength and gas barrier ability of cellulose films and sheets can be increased by promoting crosslinking between fibril units. For example, cellulose sheets undergone through UV radiation showed increased tensile strength since UV radiation enables intercross linking between the neighboring cellulose molecules [23].

Homogenization and related mechanical processes including spinning, flow-focusing, microfluidization, and ultrasonication can modify the structure of cellulose fibrils into microfibrils and nanofibrils through mechanical breakdown. Cellulose filaments are typically continuous fibers of random length. Cellulose microfibrils, on the other hand, are smaller fiber agglomerates with diameters ranging from 3.5 nm to 35 nm. Nanofibrils are smaller than microfibrils in width nanometric range and the length of micrometer range. Cellulose microfibrils store three different domains, namely, non-crystalline, paracrystalline and crystalline regions. The former chains include regions without regular packing of cellulose chains. The paracrystalline domain has partial crystal distortion and loose molecular packing. As its name implies, the crystalline region takes shape with highly ordered chains. Furthermore, additional networks such as hydrogen-bonded bundles or microfibrils can be observed on microfibril surfaces due to hydroxyl groups. Cellulose micro or nanofibril units possess large surface area and strength. Hence, films made from cellulose micro/nano fibrils are known to have low thermal expansion coefficient, high transparency and good mechanical characteristics [24]. Cellulose nanofibril films may even show better oxygen and water vapor barrier ability than some petroleum-based polymers [25]. Furthermore, mechanical properties are important for handling and packaging applications to interpret the durability and sustainability of materials. Cellulose nanofibrils mixed with other polymers can exhibit good mechanical features such as high elasticity and tensile strength [26]. Application of high-shear and high-pressure processes (e.g., homogenization and microfluidization) on cellulose dispersions dates back to couple of decades. These techniques can be used during extraction to obtain microfibers. For example, They can provide sufficient mechanical force to breakdown native cellulose fibers in the pulp into microfibrillated cellulose gels [27]. Another research demonstrated the use of microfluidization in combination with enzyme treatment to obtain nanofibrillated cellulose from palm fruit peels [28, 29]. Extrusion and flow spinning were used in a similar way to modify cellulose fibers in their solution, which is extruded through a spinneret and rotating godets with a certain velocity and pressure. The cellulose solution pushed through the axial direction, generates a highly oriented form. These cellulose filaments contain highly crystalline and dense microfibrillar structure leading to strong and moisture resistant film materials [30].

3.2 Chemical modifications

3.2.1 Hydrolysis

Hydrolysis involves partial depolymerization of cellulose by using dilute mineral acid. Microcrystalline cellulose (MCC) constitutes the micron scale highly crystalline regions of cellulose microfibrils. MCC can be generated by the hydrolysis of partially depolymerized α-cellulose using hydrochloric acid. MCC is preferred as a biocomposite in various application areas, such as medicine and pharmacology due to its biodegradability, physicochemical stability, low density, elasticity, and large surface area. MCC was first discovered and commercialized as Avicel® since early twentieth century. Commercial production is conventionally from softwood, hardwood, cotton stalks, soybean husks or rice husks. MCC dispersions are decomposed using sulfuric acid to obtain hydrophobic or hydrophilic CNC depending on the solvent polarity. In parallel to MCC, nanocrystalline cellulose (or cellulose nanocrystals, CNC) are rod-shaped highly crystalline units of cellulose microfibrils at nanoscale dimensions. Processes similar to MCC are involved in production of CNC by hydrolysis of native cellulose commonly from algae, bacteria and cotton linter. CNC enjoys optical transparency, light weight, biocompatibility and low thermal expansion for uses in biomedical and packaging applications. There are examples of manufacturing packaging films from CNC using layer by layer deposition or spray drying techniques. The former method is based upon adsorption of the oppositely charged polyelectrolytes to create uniform nanocellulose multilayers [31].

3.2.2 Esterification

Esterification or acylation process involves the combination of organic acid and alcohol to form an ester and water. Esterification reactions can occur on cellulose polymer chains or the outer surface of cellulose fibers by aromatic and aliphatic reagents. Cellulose esters can be classified as organic (e.g., cellulose acetate, cellulose acetate butyrate and cellulose acetate propionate) and inorganic (cellulose nitrate, cellulose phosphate, cellulose sulfate, and cellulose xanthate) esters.

Cellulose acetate is a well-known and one of the most versatile cellulose derivatives. It can be obtained by the reaction of cellulose with acetic acid, acetic anhydride and sulfuric acid as catalysts. Degree of substitution with the acetate groups is the average number of acetyl groups replacing the hydroxyl groups per glucose unit, and it determines the solubility of cellulose acetate in water. The maximum degree of acetylation is three, when all OH groups are replaced by acetyl groups, which has the highest water solubility. Cellulose acetate has a great commercial appeal due to its versatility related to its solubility in various solvents and toughness for mechanical strength. Therefore, it found applications in distinct industry segments which require either flexible and resistant material as in photography films or rigid and durable materials produced by molding such as in glass frames, combs, tool handles, solvent reservoirs, etc. Therefore, it is not surprising to find packaging applications of cellulose acetate due its high chemical and thermal stabilities, dense and strong structure, and low cost [32]. The versatility of cellulose acetate allows its adoption to different film production technologies [6]. Electrospinning was used a promising and affordable method to manufacture packaging material. This technique relies on the application of an electrical potential to create thin cellulose fibers extruded through a nozzle [33].

Cellulose sulfate is an inorganic derivative produced with the addition of sulfuric acid, sulfur trioxide or chlorinated sulfonic acid to cellulose fiber. Esterification can occur via two strategies. Heterogeneous sulfation includes SO_3 complexes such as DMF as solvent, pyridine as sulfation agent and H_2SO_4 as the reactant. The sulfation of cellulose results in chain cleavage and produces cellulose sulfate with a high degree of substitution. Cellulose sulfate produced with heterogeneous method may have non-uniform distribution of crystalline and amorphous regions and limited solubility in water. In contrast, homogeneous sulfation requires N_2O_4-DMF and SO_3 complexes for the synthesis of cellulose sulfate. It generates even distribution of crystalline and amorphous regions, which results in higher solubility in water and low-chain degradation. When used in packaging manufacturing, cellulose sulfate can represent antimicrobial characteristics. In addition, cellulose sulfate films are recognized as biodegradable and biocompatible, different from cellulose acetate.

Cellulose phosphate is an inorganic cellulose derivative generated by phosphorylation of cellulose fibers using phosphor-containing esters such as phosphonites, phosphites, or phosphinates. Similar to sulfation, phosphorylation may take place under heterogeneous or homogeneous conditions. Heterogeneous phosphorylation of cellulose fiber occurs in the presence of solvents such as *N*-methylmorpholine *N*-oxide or DMF. The homogeneous approach results in complete or partial esterification of cellulose fiber to yield a soluble product. Unlike cellulose sulfate, cellulose phosphate is largely water-insoluble due to polymer chain crosslinking. This may result in coagulation in the solution and serves as a limitation for applications that require a free-flowing melt, such as packaging [7].

Cellulose nitrate is the poly-nitrate ester of cellulose obtained by replacing hydroxyl groups with nitrates using nitrating acid mixtures. Cellulose nitrate demonstrates odorless and tasteless characteristics and is typically soluble in organic solvents (e.g., alcohols), which allow it to be used in various areas, such as packaging film, membrane filters, pharmaceutical, and optical instruments. The degree of nitration is an important parameter for cellulose nitrate that shows the volume of nitric oxide release when 1 g of cellulose nitrate is completely decomposed at 0°C and 1 atm. It influences the physicochemical properties of cellulose nitrate such as viscosity, solubility, heat of formation and combustion and thermal stability [34, 35]. Generally, lower degree of nitration (10–11%) is preferred for film manufacturing as it provides thermoplastic behavior and solubility in organic solvents, which are essential for film manufacturing [36].

Cellulose xanthate is a sodium hydroxide-soluble cellulose derivative. The control over the process is critical since industrial viscose rayon fabric (known as viscose) and cellophane are formed from xanthate. It is obtained by stepwise xanthation reactions, which involve alkalization using sodium hydroxide, followed by ripening to bring the degree of polymerization to 0.5 (i.e., to adjust the viscosity). After ripening, the solution is passed through a spinneret to create soft filaments [37]. Finally, the soft filaments are exposed to sulfuric acid and zinc, and sodium sulfate to obtain the rayon fiber filaments. Rayon fabric shows good plastic behavior with high mechanical strength, elasticity and strong moisture absorption. Degree of ripening and polymerization determine the strength of rayon fibers: higher DP and DR increased the strength of the films prepared by fiber spinning [38]. The high moisture absorption rate is related to their lower crystallinity that serves as a drawback for packaging films applications. Rayon fabric is commercially produced from wood pulp. The physicochemical features of the rayon fabric are close to fabrics produced from natural cellulose (e.g., cotton or linen). The viscose fibers show high resistance to acid and alkali conditions, and resistant to fabric and film manufacturing conditions [39]. Versatility and affordability allow viscose for its use in large-scale industrial textiles and food packaging, such as sausage casings [40].

3.2.3 Etherification

Etherification of cellulose involves homogeneous or heterogeneous reactions with etherifying agents, such as epoxides, halogenated alkanes, alkyl, silyl chlorides, bromides, or vinyl compounds. Cellulose ethers are classified as silyl ethers, ionic alkyl ethers and nonionic alkyl ethers and have alkyl halide structures with hydroxyl groups. They are among the most abundant cellulose derivatives and manufactured on large scales. The commercial forms used in various industries include methylcellulose, carboxymethyl cellulose, ethyl cellulose, hydroxyethyl cellulose, and hydroxypropyl cellulose. Cellulose ethers represent non-toxic, tasteless and non-flammable and water-soluble features. These properties enable them to be used as a thickener, water binder, processing aids in food formulations; as excipients in pharmaceutical formulations; and textile and packaging applications [41]. Their solubility depends on the pH of the solution: soluble in alkaline conditions with decrease in solubility with decreasing pH. They are also thermally stable up to 100°C, and form crosslinking network with slight degradation at 130–150°C [42].

Methyl cellulose is the nonionic alkyl ether that is produced with methyl chloride or dimethyl sulfate in an alkaline medium. Methyl cellulose demonstrates high solubility in water and organic solvents, good gelling capabilities and emulsifying power. Therefore, they are common ingredients in sugar-based syrups, oil emulsions, creams, food gels, baked goods, fried foods, and soups. Moreover, methyl cellulose has a potential in food packaging film industry as a plasticizer since they display low cost, transparency, high strength, easy handling and environmentally friendly structure that are important properties for packaging films [43].

Carboxymethyl cellulose (CMC) is an anionic cellulose ether derivative obtained by the reaction between alkali cellulose and sodium chloroacetate, where hydroxyl groups are replaced with carboxymethyls at 2nd, 3rd, and 6th carbon of a glucose unit. CMC is another traditional food additive as a thickener, stabilizer, emulsifier, bulking agent, and water binder related to its affordability, water solubility, chemical stability, and non-toxicity. It is a common ingredient in food gels, such as processed meat products, dry pet foods, yoghurt, beverages, ice cream and dessert. In addition to these, CMC can provide control to the viscosity of film-forming solution, and protective barrier when used in food packaging film applications [44].

Ethyl cellulose is a nonionic cellulose derivative obtained by the reaction between alkali cellulose and ethyl chloride between 90 and 150°C, where hydroxyl groups are converted into ethyl ether groups. This results in less polarity than CMC and methyl cellulose. Ethyl cellulose is soluble in various organic solvents such as hydrocarbons, alcohols, esters and chlorinated solvents. The physicochemical properties and applicability of ethyl cellulose depend on the degree of etherification. The conversion of hydroxyl groups into ethyl ether units provides water-solubility, adhesive features as well [45].

Hydroxyethyl cellulose is a nonionic derivative obtained by the reaction of cellulose with sodium hydroxide and ethylene oxide, where ethyl groups are attached to hydroxyl units. They show non-toxic, compressible, and water-soluble characteristics attractive for tissue engineering and biomedical applications. They are also widely used as a thickening, emulsifying agent and stabilizer in personal care and cosmetic products. It is not permitted to be used in food formulations due to its chemical structure, but it might be an alternative for food packaging films [46].

Hydroxypropyl cellulose is an alkali cellulose derivative obtained with the substitution of 2-hydroxypropyl chloride with hydroxyl groups. Similar to methyl cellulose, it shows gelling ability, however, at higher temperatures. Hydroxypropyl cellulose can be used as gelling agent, emulsifier or thickener in salad dressing, sorbet, and confectionary product fillings. Hydroxypropyl cellulose can also contribute to gas barrier properties oil-resistance of packaging films [47, 48].

3.2.4 Silylation

Silylation is based on incorporation of multifunctional silane compounds (i.e., methyltrimethoxysilane, aminopropyltriethoxysilane, methacrylopropyltrimethoxysilane alkoxysilane) into cellulose fibers or micro- and nano-crystals. Silyation provides unique characteristics depending on the specific groups in the glucose unit (i.e., −OH, −CH or −COOH). For example, cellulose nanocrystals can be partially silylated using n-dodecyldimethylchlorosilane as a silylating agent and acetone as the solvent to allow surface modification. The silylation process can effectively enhance crystallinity, crosslinking, and strength of cellulose fibers. For example, partially silylated cellulose nanocrystals exhibited improved degree of crystallinity and tensile strength [49]. Similar to other cellulose derivatives, when used in film formulations, silylation can improve their functional properties, such as water vapor barrier and thermal resistance [50].

3.2.5 Carbamylation

Carbamylation of cellulose occurs by the reaction of hydroxyl group of cellulose with isocyanate. Carbamylation of nanofibers can increase their reactivity and thermal stability. The carbamylation of cellulose modifies the polarity of cellulose fibers. For example, hydrophobicity of cellulose nanocrystals increased by crosslinking phenyl isocyanate with hydroxyl groups, while crystallinity and other physicochemical characteristics of cellulose nanofibers remained unchanged [51].

A major limitation in food and packaging applications is related to the use of isocyanate that can create toxic byproducts during the carbamylation process.

3.2.6 TEMPO oxidation

The free radical TEMPO (2,2,6,6-tetramethylpiperidine-1-oxyl) is a water soluble and relatively stable nitroxyl radical (i.e., half-life in minutes at room temperature). It can catalyze the oxidation of primary hydroxyl groups of cellulose fibrils to carboxyl form. TEMPO reaction requires oxidizers such as NaClO or $NaClO_2$ in the presence of NaBr at alkaline conditions. This process can provide electrostatic repulsion between cellulose fibers by preventing hydrogen bonding. TEMPO-oxidized cellulose fibers can be used to produce films with improved network structure, optical features, gas barrier and thermomechanical characteristics. For instance, the films prepared from TEMPO-oxidized wood cellulose nanofibers exhibited improved degree of fibrillation and low oxygen permeability, desirable optical transparency, and high tensile strength and Young's modulus in their films [52]. Similarly, softwood and hardwood celluloses oxidized by TEMPO displayed transparent, flexible and low thermal expansion coefficient with higher crystallinity compared with untreated cellulose nanofiber films [53].

4. Cellulose network and cellulose interactions

4.1 Cellulose fibrillar network

The cellulose microfibrils and their derivatives are responsible for the formation of a cellulose polymer network as seen in film structures, where they interact with other polymeric materials and doped nanoparticles through electrostatic forces, hydrogen bonds, Van der Waals attraction as illustrated in **Figure 2**. The nature of these interactions influences the strength and morphology of cellulose-based films. For example, the presence of similar charged substitutes on fiber surfaces may create repulsive forces to result in porous structure while opposite charged groups on fibril units might cause dense and smooth film surfaces [54]. In addition to this, high dispersibility of cellulose fibrillar matrices and strong hydrogen bonds with other polar moieties through hydroxyl groups are important for the strength of the cellulose-based films. For example, nanocellulose films with high fibrillar network and extensive hydrogen bonding demonstrate low porosity and high oxygen and moisture barrier [55]. Overall, these interactions support a good film framework important for desirable mechanical and gas barrier features. This section explains and exemplifies the nature and importance of cellulose interactions with a focus on physicochemical and mechanical properties of film materials.

4.2 Cellulose-protein interactions

Native cellulose is a nonionic polymer with reduced interaction capacity without pretreatment. Cellulose derivatives can show interfacial activity and electrostatic

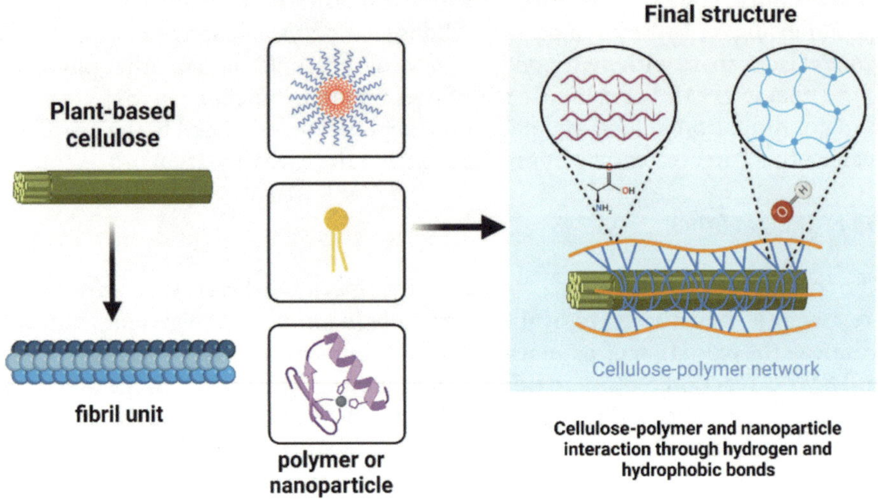

Figure 2.
A schematic illustration of cellulose interactions with polymers and nanoparticles.

potential to interact with proteins via electrostatic interaction, hydrogen bonding or hydrophobic attraction. Electrostatic interactions between protein and cellulose derivatives (e.g., negatively charged methylcellulose and carboxymethyl cellulose) can provide advanced control over the characteristics of cellulose films to improve their mechanical and barrier properties. For example, a recent research showed that cellulose nanocrystals from a bacterial source can participate in protein gels to improve mechanical properties (tensile strength) and as well as oxygen and moisture barrier abilities of nanocomposite biodegradable packaging films [56]. Another study showed that the intermolecular interactions between fish gelatin and crystalline regions of cellulose microfibrils significantly increased the tensile strength and reduced the water vapor permeability of biodegradable fish gelatin films [57]. In addition to this, the physicochemical properties (e.g., pH, temperature, ionic strength, concentration, etc.) of solutions of cellulosic polymers and proteins determine the extent of their interactions. Among these parameters, pH played the most critical role since it affects the surface charges, wettability, density and conformational state of dispersed systems [58]. For example, at lower pH (pH 3), soy protein isolate (SPI) and TEMPO oxidized bacterial cellulose interacted via electrostatic attractions and formed a positively charged noncovalent complex that displayed good creaming stability and elastic gel texture. On the other hand, at higher pH (pH 9), cellulose-protein interaction shifted from strong electrostatic attraction to repulsion [59]. Moreover, at pH 4 negatively charged CMC can form a complex with positively charged pea protein to form insoluble complexes. At lower pH values, the interaction between CMC and protein increases with strong electrostatic repulsion, high viscosity, and steric hindrance [60]. In addition, the presence of small molecules, such as salts or sugars, may alter the nature of interactions. For example, ionic liquids, which are molten salts with organic cations and anions, show significant effects on cellulose-protein complexes [61]. Anions may form hydrogen bonds with hydroxyl groups found in cellulose or protein and disrupt the naturally occurring hydrogen bond between these two polymers. Moreover, cations in ionic liquids may interact with ether oxygen atoms or CH groups in native cellulose or derivatives [62]. It was shown that chloride ion associated with hydroxyl groups in cellulose and resulted in weakening intermolecular hydrogen bonds between protein-cellulose complex extracted from peanut leaf. These interactions reduced the tensile strength of prepared films [63].

4.3 Cellulose-polysaccharide interactions

The interactions between cellulose and polysaccharides, such as starch, glucomannan, pullulan occur between unsubstituted chains via hydrogen bonding. Anionic cellulose derivatives, such as CMC and CNC can interact with charged polysaccharides, such as sodium alginate or chitosan to form hydrogel structures by intermolecular inclusion interaction. These interactions occur via hydrogen bond or ionic crosslinking and enhances mechanical performances For example, chitosan-carboxymethyl cellulose interaction creates polyelectrolyte multilayer films that includes oppositely charged layers of polyelectrolytes [64]. The electrostatic interaction between these two polymers results in binding of the oppositely charged polymers on the film surface. This brings about improved strength in composite film. Another study showed that intermolecular interaction between konjac glucomannan and pullulan and natural cellulose nanofibrils yields strong crosslinks to improve the flexibility and elasticity of films and reduce the water vapor permeability [65]. Besides, strong hydrogen bonds between hydroxyl groups of cellulose nanofibrils and hydroxyl or carboxyl groups of sodium alginate increased the cohesiveness and water resistance of the biopolymer films [66]. The authors in the reference [67] examined the influence of molecular interactions between microcrystalline cellulose (MCC) and propylene glycol alginate-agar polymeric mixture and their effect on mechanical, physical and barrier properties of microcrystalline cellulose gum edible films. The addition of 4% MCC was sufficient to reduce the water vapor permeability and enhance the tensile strength of the film due to inter and intramolecular hydrogen bonds. Interestingly, incorporation of cellulose resulted in reduced intramolecular interactions between polymers and caused less compact network and hydrophobicity on the film surfaces. A study focused on the effect of electrostatic interactions between cellulose nanofiber and alginate or chitosan in their films [68]. It was highlighted that cellulose-polysaccharide interaction mechanisms led to an increase in tensile strength and water vapor resistance of the polysaccharide films. Another study modified CMC films with xanthan gum, and flaxseed gum to improve the physical and mechanical properties (e.g., water vapor resistance, tensile strength, elasticity), and reduce the weight loss and increase shelf life of mango [69].

4.4 Cellulose-nanoparticle interactions

Packaging films prepared from native cellulose or derivatives, such as cellulose nanocrystals or CMC, can be doped with nanoparticles to provide them strength, stability, ultraviolet barrier, optical or antimicrobial functionality. For example, paramagnetic iron oxide nanoparticles coating to cellulose nanocrystals was shown to improve the thermostability of the films [70]. Another study focused on the preparation of cellulose nanocrystal films incorporated silver (Ag) [71]. The films showed ultraviolet barrier property and reduced water vapor permeation. Furthermore, calcium hydroxide nanoparticles can support the crosslinking between polysaccharide and cellulose nanofiber by diffusing into the matrix and attaching with ionic bonds. This enables to increase in the opacity and thermal stability of multilayer packaging films. Photocatalytic TiO_2–Ag nanoparticles can be added to provide antimicrobial and photocatalytic activity to CMC based films. In addition, they also increased the tensile strength of the films related to increased electrostatic attractions between hydroxyl groups and O−Ti−O bonds between CMC and nanoparticles [72]. In another study, ZnO-loaded cellulose acetate film was obtained by ionic interaction between zinc and oxygen atoms in cellulose acetate [73]. These films exhibited antibacterial action against *E. coli* and moisture barrier characteristics. Recently, Amini et al. [74] created nano-composite packaging films using cotton cellulose and hydrophobic polycaprolactone (PCL) dissolved in an ionic liquid containing zinc oxide nanoparticles. The intermolecular hydrogen bond between cellulose fibers, PCL and zinc oxide resulted in promising film properties (i.e., improved oxygen barrier property, good tensile energy absorption, elasticity, and surface smoothness). The authors mentioned that nanocomposite film may serve as a packaging film for peanuts and meat based on physical and mechanical characteristics. An interesting study mentioned that CMC films prepared with the addition of cobalt nanosheets show strong antimicrobial activity against *E. coli and S.s aureus* without affecting mechanical properties of the films [75].

5. Conclusion and future perspectives

In conclusion, several literature and industrial examples indicate great potential for cellulose and its derivatives to be used in manufacturing of film materials with tailored properties for specific applications and needs. Cellulose and derivatives are potential sources for commercial packaging films due to their good film-forming abilities. There is an increasing trend in cellulose-based biodegradable packaging applications in combination with active ingredients with antioxidant and antimicrobial properties. Cellulose fibers are abundant and can be obtained from affordable and sustainable resources extracted from plants, animals, bacterial or algae-based sources and as well as waste streams of industrial production lines. The research is still ongoing to overcome certain challenges and improve film properties in consideration of environmental sustainability, biodegradability as well as mechanical and gas permeation resistance properties of the film materials. Controlling the micro-scale interactions between cellulose and other polymers as well doped nanoparticles can be used to strengthen the film network and furnish them with new functionalities. As a general trend of following research efforts, development and scaling up the emergent technologies to large scale production and industrial applications remain as one of the biggest challenges. For example, industrial production of films requires specialized instruments where the materials need to be resistant to high temperatures and pressures associated with the extruders. Overall, advances in cellulose modification methods can address practical challenges associated with adoption of their films as future perspectives.

DOI: http://dx.doi.org/10.5772/intechopen.110817

Author details

Eda Ceren Kaya[1] and Umut Yucel[2]*

1 Food Science Institute, Kansas State University, Manhattan, KS, USA

2 Food Science Institute, Animal Sciences and Industry Department, Kansas State University, Manhattan, KS, USA

*Address all correspondence to: yucel@ksu.edu

References

[1] Jang M, Shim WJ, Cho Y, et al. A close relationship between microplastic contamination and coastal area use pattern. Water Research. 2020;**171**:115400

[2] Saravanan A, Karishma S, Kumar PS, et al. Biodegradation of oil-contaminated aqueous ecosystem using an immobilized fungi biomass and kinetic study. Environmental Research. 2023;**220**:115252

[3] Abdullah CJ, Hafeez MA, et al. Biopolymer-based functional films for packaging applications: A review. Frontiers in Nutrition. 2022;**9**:1-20

[4] Shrestha B, Dhungana PK, Dhital S, et al. Evaluation of modified Sorghum starches and biodegradable films. Journal of Food Science Technology Nepal. 2018;**10**:11-17

[5] Song T, Qian S, Lan T, et al. Recent advances in bio-based smart active packaging materials. Foods. 2022;**2022**:11. DOI: 10.3390/foods11152228

[6] Liu Y, Ahmed S, Sameen DE, et al. A review of cellulose and its derivatives in biopolymer-based for food packaging application. Trends in Food Science and Technology. 2021;**112**:532-546

[7] Onwukamike KN, Grelier S, Grau E, et al. Critical review on sustainable homogeneous cellulose modification: Why renewability is not enough. ACS Sustainable Chemistry & Engineering. 2019;7:1826-1840

[8] Saedi S, Garcia CV, Kim JT, Shin GH. Physical and chemical modifications of cellulose fibers for food packaging applications. Cellulose. 2021;**28**:8877-8897

[9] Henriksson M, Berglund LA, Isaksson P, et al. Cellulose nanopaper structures of high toughness. Biomacromolecules. 2008;**9**:1579-1585

[10] Padrão J, Gonçalves S, Silva JP, et al. Bacterial cellulose-lactoferrin as an antimicrobial edible packaging. Food Hydrocolloids. 2016;**58**:126-140

[11] Keshk SMAS. Vitamin C enhances bacterial cellulose production in Gluconacetobacter xylinus. Carbohydrate Polymers. 2014;**99**:98-100

[12] Shankar S, Rhim J. Preparation of nanocellulose from micro-crystalline cellulose : The effect on the performance and properties of agar-based composite films. Carbohydrate Polymers. 2016;**135**:18-26

[13] Cebrián-Lloret V, Metz M, Martínez-Abad A, et al. Valorization of alginate-extracted seaweed biomass for the development of cellulose-based packaging films. Algal Research. 2022;**2022**:61. DOI: 10.1016/j.algal.2021.102576

[14] Zhao Y, Zhang Y, Lindström ME, et al. Tunicate cellulose nanocrystals: Preparation, neat films and nanocomposite films with glucomannans. Carbohydrate Polymers. 2015;**117**:286-296

[15] Xia J, Zhang Z, Liu W, et al. Highly transparent 100% cellulose nanofibril films with extremely high oxygen barriers in high relative humidity. Cellulose. 2018;**25**:4057-4066

[16] Chen YM, Wan JQ, Huang MZ, et al. Influence of drying temperature and duration on fiber properties

of unbleached wheat straw pulp. Carbohydrate Polymers. 2011;**85**:759-764

[17] Panaitescu DM, Vizireanu S, Nicolae CA, et al. Treatment of nanocellulose by submerged liquid plasma for surface functionalization. Nanomaterials. 2018;**2018**:8. DOI: 10.3390/nano8070467

[18] Syverud K, Stenius P. Strength and barrier properties of MFC films. Cellulose. 2009;**16**:75-85

[19] Ragoubi M, Bienaimé D, Molina S, et al. Impact of corona treated hemp fibres onto mechanical properties of polypropylene composites made thereof. Industrial Crops and Products. 2010;**31**:344-349

[20] Ragoubi M, George B, Molina S, et al. Effect of corona discharge treatment on mechanical and thermal properties of composites based on miscanthus fibres and polylactic acid or polypropylene matrix. Composites. Part A, Applied Science and Manufacturing. 2012;**43**:675-685

[21] Moise IV, Manea MM, Vasilca S, et al. The crosslinking behaviour of cellulose in gamma irradiated paper. Polymer Degradation and Stability. 2019;**160**:53-59

[22] Khan RA, Beck S, Dussault D, et al. Mechanical and barrier properties of nanocrystalline cellulose reinforced poly(caprolactone) composites: Effect of gamma radiation. Journal of Applied Polymer Science. 2013;**129**:3038-3046

[23] Zaman HU, Khan MA, Khan RA. Improvement of mechanical properties of jute fibers-polyethylene/polypropylene composites: Effect of green dye and UV radiation. Polymer—Plastic Technology Engineering. 2009;**48**:1130-1138

[24] Capadona JR, Berg O, Capadona LA, et al. A versatile approach for the processing of polymer nanocomposites with self-assembled nanofibre templates. Letters. 2007;**2007**:765-769

[25] Mazega A, Tarr Q, Aguado R, et al. Improving the barrier properties of paper to moisture , air , and grease with nanocellulose-based coating suspensions. Nanomaterials. 2022;**2022**:1-14

[26] Sessini V, Navarro-Baena I, Arrieta MP, et al. Effect of the addition of polyester-grafted-cellulose nanocrystals on the shape memory properties of biodegradable PLA/PCL nanocomposites. Polymer Degradation and Stability. 2018;**152**:126-138

[27] Dinand E, Chanzy H, Vignon MR. Suspensions of cellulose microfibrils from sugar beet pulp. Food Hydrocolloids. 1999;**13**:275-283

[28] Ferrer A, Filpponen I, Rodríguez A, et al. Valorization of residual Empty Palm Fruit Bunch Fibers (EPFBF) by microfluidization: Production of nanofibrillated cellulose and EPFBF nanopaper. Bioresource Technology. 2012;**125**:249-255

[29] Tibolla H, Pelissari FM, Rodrigues MI, et al. Cellulose nanofibers produced from banana peel by enzymatic treatment: Study of process conditions. Industrial Crops and Products. 2017;**95**:664-674

[30] Peresin MS, Habibi Y, Vesterinen AH, et al. Effect of moisture on electrospun nanofiber composites of poly(vinyl alcohol) and cellulose nanocrystals. Biomacromolecules. 2010;**11**:2471-2477

[31] Rojas-Lema S, Terol J, Fages E, et al. Microencapsulation of copper(ii) sulfate in ionically cross-linked chitosan by spray drying for the

development of irreversible moisture indicators in paper packaging. Polymers (Basel). 2020;**2020**:12. DOI: 10.3390/POLYM12092039

[32] Liang L, Chen H. Development and characterization of biodegradable ultraviolet protective and antibacterial polylactic acid-cellulose acetate film modified by phenyl salicylate. International Journal of Biological Macromolecules. 2022;**211**:85-93

[33] Attari N, Hausler R. Reinforcing effects of fibrous and crystalline nanocelluloses on cellulose acetate membranes. Carbohydrate Polymer Technology Application. 2023;**5**:100281

[34] Tarchoun AF, Trache D, Klapötke TM, et al. Chemical design and characterization of cellulosic derivatives containing high-nitrogen functional groups: Towards the next generation of energetic biopolymers. Def Technology. 2022;**18**:537-546

[35] Zhang S, Mi Y, Jin J, et al. Research on ignition characteristics of nitrocellulose in warehouse. In: 2019 9th Int Conf Fire Sci Fire Prot Eng ICFSFPE 2019. 2019. DOI: 10.1109/ICFSFPE48751.2019.9055813

[36] Fodil Cherif M, Trache D, Benaliouche F, et al. Organosolv lignins as new stabilizers for cellulose nitrate: Thermal behavior and stability assessment. International Journal of Biological Macromolecules. 2020;**164**:794-807

[37] Zhou S, Li Y, Huang L, et al. Enhanced reactivity of Kraft-based viscose-grade dissolving pulp by hollander beating treatment. ACS Sustainable Chemistry & Engineering. 2018;**13**:2861-2870

[38] Worku LA, Bachheti RK, Tadesse MG. Isolation and characterization of natural cellulose from *Oxytenanthera abyssinica* (Lowland Ethiopian Bamboo) using alkali peroxide bleaching stages followed by aqueous chlorite in buffer solution. International Journal of Polymer Science. 2022;**2022**. DOI: 10.1155/2022/5155552

[39] Saleem MA, Pei L, Saleem MF, et al. Sustainable dyeing of nylon fabric with acid dyes in decamethylcyclopentasiloxane (D5) solvent for improving dye uptake and reducing raw material consumption. Journal of Cleaner Production. 2021;**279**:123480

[40] Du J, Li C, Zhao Y, et al. Hemicellulose isolated from waste liquor of viscose fiber mill for preparation of polyacrylamide-hemicellulose hybrid films. International Journal of Biological Macromolecules. 2018;**108**:1255-1260

[41] Ostrovskii D, Kjøniksen AL, Nyström B, et al. Association and thermal gelation in aqueous mixtures of ethyl(hydroxyethyl)cellulose and ionic surfactant: FTIR and Raman study. Macromolecules. 1999;**32**:1534-1540

[42] McMullen RL, Ozkan S, Gillece T. Physicochemical properties of cellulose ethers. Cosmetics. 2022;**2022**:9. DOI: 10.3390/cosmetics9030052

[43] López De Dicastillo C, Rodríguez F, Guarda A, et al. Antioxidant films based on cross-linked methyl cellulose and native Chilean berry for food packaging applications. Carbohydrate Polymers. 2016;**136**:1052-1060

[44] Bourbon AI, Costa MJ, Maciel LC, et al. Active carboxymethylcellulose-based edible films: Influence of free and encapsulated curcumin on films' properties. Foods. 2021;**2021**:10. DOI: 10.3390/foods10071512

[45] Davidovich-Pinhas M, Barbut S, Marangoni AG. The gelation of oil using ethyl cellulose. Carbohydrate Polymers. 2015;**117**:869-878

[46] Risaliti L, Yu X, Vanti G, et al. Hydroxyethyl cellulose hydrogel for skin delivery of khellin loaded in ascosomes: Characterization, in vitro/ in vivo performance and acute toxicity. International Journal of Biological Macromolecules. 2021;**179**:217-229

[47] Giray NS, Baysal T, Taştan Ö. Effects of hydroxypropyl methylcellulose film incorporated with carvacrol and its nanoemulsion on the quality of grapes. Chiang Mai Journal of Science. 2020;**47**:441-454

[48] Song J, Feng H, Wu M, et al. Preparation and characterization of arginine-modified chitosan/ hydroxypropyl methylcellose antibacterial film. International Journal of Biological Macromolecules. 2020;**145**:750-758

[49] Zhang Z, Tingaut P, Rentsch D, et al. Controlled silylation of nanofibrillated cellulose in water: Reinforcement of a model polydimethylsiloxane network. ChemSusChem. 2015;**8**:2681-2690

[50] Jankauskaite V, Balčiunaitiene A, Alexandrova R, et al. Effect of cellulose microfiber silylation procedures on the properties and antibacterial activity of polydimethylsiloxane. Coatings. 2020;**2020**:10. DOI: 10.3390/ COATINGS10060567

[51] Missoum K, Sadocco P, Causio J, et al. Antibacterial activity and biodegradability assessment of chemically grafted nanofibrillated cellulose. Materials Science and Engineering: C. 2015;**45**:477-483

[52] Wakabayashi M, Fujisawa S, Saito T, et al. Nanocellulose film properties tunable by controlling degree of fibrillation of TEMPO-oxidized cellulose. Frontiers in Chemistry. 2020;**8**:1-9

[53] Fukuzumi H, Saito T, Iwata T, et al. Transparent and high gas barrier films of cellulose nanofibers prepared by TEMPO-mediated oxidation. Biomacromolecules. 2009;**10**:162-165

[54] Ahola S, Salmi J, Johansson LS, et al. Model films from native cellulose nanofibrils. Preparation, swelling, and surface interactions. Biomacromolecules. 2008;**9**:1273-1282

[55] Aulin C, Gällstedt M, Lindström T. Oxygen and oil barrier properties of microfibrillated cellulose films and coatings. Cellulose. 2010;**17**:559-574

[56] Leite LSF, Moreira FKV, Mattoso LHC, et al. Electrostatic interactions regulate the physical properties of gelatin-cellulose nanocrystals nanocomposite films intended for biodegradable packaging. Food Hydrocolloids. 2021;**113**:106424

[57] Santos TM, de Souza Filho M, Caceres CA, et al. Fish gelatin films as affected by cellulose whiskers and sonication. Food Hydrocolloids. 2014;**41**:113-118

[58] Zhang X, Luo X, Wang Y, et al. Concentrated O/W Pickering emulsions stabilized by soy protein/cellulose nanofibrils: Influence of pH on the emulsification performance. Food Hydrocolloids. 2020;**108**:106025

[59] Zhang X, Lei Y, Luo X, et al. Impact of pH on the interaction between soybean protein isolate and oxidized bacterial cellulose at oil-water interface: Dilatational rheological and emulsifying properties. Food Hydrocolloids. 2021;**115**:106609

[60] Klassen DR, Elmer CM, Nickerson MT. Associative phase separation involving canola protein isolate with both sulphated and carboxylated polysaccharides. Food Chemistry. 2011;**126**:1094-1101

[61] Tomimatsu Y, Suetsugu H, Yoshimura Y, et al. The solubility of cellulose in binary mixtures of ionic liquids and dimethyl sulfoxide: Influence of the anion. Journal of Molecular Liquids. 2019;**279**:120-126

[62] Blessing B, Trout C, Morales A, et al. The impact of composition and morphology on ionic conductivity of silk/cellulose bio-composites fabricated from ionic liquid and varying percentages of coagulation agents. International Journal of Molecular Sciences. 2020;**21**:1-23

[63] Zhang W, Li Y, Wang K, et al. Effect of different ionic liquids and organic solvents on the structural and physicochemical properties of cellulose-protein complexes extracted from Se-enriched peanut leaves. International Journal of Biological Macromolecules. 2022;**217**:171-179

[64] Taketa TB, Rocha Neto JBM, Dos Santos DM, et al. Tracking sulfonated polystyrene diffusion in a chitosan/carboxymethyl cellulose layer-by-layer film: Exploring the internal architecture of nanocoatings. Langmuir. 2020;**36**:4985-4994

[65] Bu N, Huang L, Cao G, et al. Konjac glucomannan/Pullulan films incorporated with cellulose nanofibrils-stabilized tea tree essential oil Pickering emulsions. Colloids Surfaces A Physicochemical Engineering Aspects. 2022;**650**:129553

[66] Deepa B, Abraham E, Pothan LA, et al. Biodegradable nanocomposite films based on sodium alginate and cellulose nanofibrils. Materials (Basel). 2016;**9**:1-11

[67] Harnkarnsujarit N, Li Y. Structure–property modification of microcrystalline cellulose film using agar and propylene glycol alginate. Journal of Applied Polymer Science. 2017;**134**:1-10

[68] Zhao K, Wang W, Teng A, et al. Using cellulose nanofibers to reinforce polysaccharide films: Blending vs layer-by-layer casting. Carbohydrate Polymers. 2020;**2020**:227. DOI: 10.1016/j.carbpol.2019.115264

[69] Zheng M, Su H, Xiao R, et al. Effects of Polygonatum cyrtonema extracts on the antioxidant ability, physical and structure properties of carboxymethyl cellulose-xanthan gum-flaxseed gum active packaging films. Food Chemistry. 2023;**403**:134320

[70] Mohammadi F, Moeeni M, Li C, et al. Interaction of cellulose and nitrodopamine coated superparamagnetic iron oxide nanoparticles with alpha-lactalbumin. RSC Advances. 2020;**10**:9704-9716

[71] Yadav M, Liu YK, Chiu FC. Fabrication of cellulose nanocrystal/silver/alginate bionanocomposite films with enhanced mec. Nanomaterials;**9**

[72] Farshchi E, Pirsa S, Roufegarinejad L, et al. Photocatalytic/biodegradable film based on carboxymethyl cellulose, modified by gelatin and TiO_2-Ag nanoparticles. Carbohydrate Polymers. 2019;**216**:189-196

[73] Chaurasia V, Chand N, Bajpai SK. Water sorption properties and antimicrobial action of zinc oxide nanoparticles-loaded cellulose acetate films. Journal of Macromolecular Science, Part A Pure and Applied Chemistry. 2010;**47**:309-317

[74] Amini E, Valls C, Roncero MB. Industrial Crops & Products Promising nanocomposites for food packaging based on cellulose—PCL films reinforced by using ZnO nanoparticles in an ionic liquid. Industrial Crops and Products. 2023;**193**:116246

[75] Tang Q, Zou Z, Huang Y, et al. Novel ammonia-responsive carboxymethyl cellulose/Co-MOF multifunctional films for real-time visual monitoring of seafood freshness. International Journal of Biological Macromolecules. 2023;**230**:123129

Chapter 6

Perspective Chapter: Cellulose in Food Production – Principles and Innovations

Hassan Sabbaghi

Abstract

For several decades, cellulose and its derivatives have been used in various fields of food processing and their applications have become increasingly important. Nutritionally, cellulose is known as dietary fiber and is used as a functional food component. Many new technological needs were recognized since developing of industrial products and, therefore, cellulose modifications (chemically or physically) also have been considered. The various important properties for using these compounds include organic solubility, gel and film formation ability, make mucoadhesive system, high swelling, hydrophilic and hydrophobic features, act as viscosifying agent, and thermoplastic effects in food systems. Thus, the most typical technological role of these cellulose's characteristics can be mentioned as edible coating (in fruits and fried products), edible film, emulsification, stabilizing agent, rheology control, suspending agent, diffusion barrier, encapsulation, extrusion, molding, and foam stabilizer in food industry. The new innovations can be mentioned as the production of bacterial cellulose, developing the smart packaging, and the preparation of nano-cellulose with environmentally friendly processes. Finally, with the expansion of the usage of cellulosic materials, a reevaluation of their food safety has been carried out. Also, the legal guidelines related to the use of these compounds as raw materials have been provided for manufacturers.

Keywords: cellulose, derivatives, technological role, food industry, innovations

1. Introduction

Cellulose is the most considerable biopolymer (naturally occurring polymer) on the ground with 1.5×10^{12} heaps of annual manufacturing. This organic polymer is most regularly obtained from plants, and nowadays its significance in food applications and personal and scientific care cannot be dominated out. Cellulose is additionally obtained *via* in vitro synthesis with the assistance of enzymatic pathways, the chemical synthesis from glucose derivatives, and the biosynthesis *via* various micro-organisms, such as algae and fungi, as nicely as a number of cardio nonpathogenic bacteria of the genera Agrobacterium, Sarcina, Rhizobium, and Acetobacter [1]. Also, cellulose can be modified using chemical reactions to manufacture some compounds with special characteristics. The changes in cellulose polymer enhance and develop its

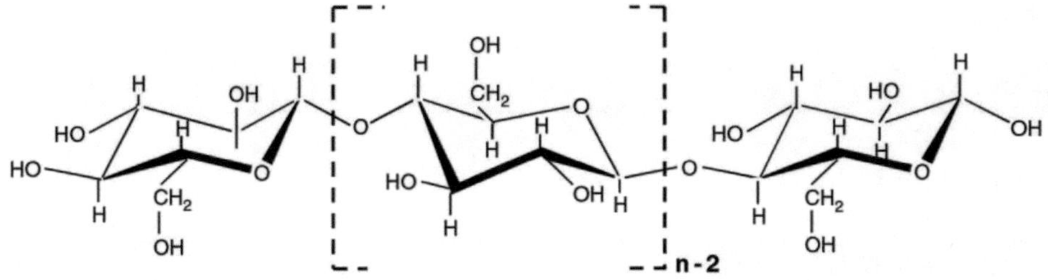

Figure 1.
Linear molecule of cellulose polymer with D-glucose units.

chemical and physical facets, which allow the usage of derivatives for many purposes and applications [2]. Structurally according to **Figure 1**, cellulose is a polysaccharide that is composed of quite a few lots to over ten hundred D-glucose units as a linear chain [3].

Cellulose is recognized as "Dietary fiber." Dietary fiber includes remains of plant cells resistant to hydrolysis as digestion procedure via the alimentary enzymes of gastrointestinal system in the human body, whose components are cellulose, hemicellulose, lignin, waxes, pectins, gums, and mucilages. These compounds are naturally current in nuts, cereals, fruits, and vegetables. The quantity and composition of fibers vary from meal to meal based on food structure. A fiber-rich diet is less in calorie density, frequently has a limited fat content, is large in extent, and is richer in micronutrients. This large mass of meals takes longer to consume and its presence in the belly might also deliver a feeling of satiety sooner, although this feeling of fullness is quick time period. It is advised that healthy adults need to devour between 20 and 35 g of dietary fiber every day [4]. There are two kinds of dietary fiber: soluble and insoluble. The insoluble substances structure a bulky mass and velocity transit time through the gastrointestinal tract due to the fact of their bulk; cellulose, hemicellulose, lignin, and waxes fall into this category. Among the soluble dietary fibers are the pectins, gums, and mucilages. All of these have the capability to hold water and thereby expand the viscosity of the food mixtures [5]. Cellulose polymer and its various derivatives have long been applied in manufacturing formulated foods. The modified celluloses (chemical or physical) are beneficial in many food products and the place bulk properties are acceptable targets. This would consist of healthy products, reduced or low-fat foods, taste oil imbibers, or flowable products such as synthetic sweeteners and taste packets. The use of these cellulosic compounds in food processing is typically due to their rheology, managed water interaction, textural attributes, and no longer to solubility or different chemical properties [6]. In this chapter, the use of cellulose biopolymer and its various derived compounds are investigated in the food industry with regard to the principles and innovations.

2. Structure modification of cellulose

Cellulose is a suitable polymer in terms of its potential to create chemical and physical modifications. Considering that the variety of food products needs to provide certain conditions of cellulose behavior and eliminate deficiencies of efficiency, the modified structures of cellulose are widely used in the food industry. Applied biopolymer is derived from cellulose that enables beneficial functions or allows really useful features in various food systems [7]. The list of commercial compounds derived from cellulose as an additive in the food and pharmaceutical industries is as follows:

- Methyl Cellulose (MC): It is produced through the substitution of hydroxyl groups on the polysaccharide molecule with methoxyl groups from the reaction between cellulose in an alkaline medium with a methylating agent, such as methyl iodide, methyl chloride, or dimethyl sulfate. Some features that make it suitable for food use are tough material, absolutely nontoxic, tasteless, odorless, and high-viscosity solutions at very low concentrations [8].

- Hydroxypropyl Methyl Cellulose (HPMC): It is manufactured by the replacement of hydroxyl groups on the polysaccharide molecule with methoxyl and hydroxypropyl groups [9]. During the production of HPMC, the methylcellulose is then further reacted with the staged addition of an alkylene oxide as propylene oxide. By making these changes, some desired modifications occur on the cellulose polymer, which should be noted. HPMC gelling temperature in hot water is considerably greater than that of MC [10]. In contrast with MC, the dissolution in cold water has additionally been noticeably improved. The viscosity of HPMC is much less affected by means of the temperature than MC, and its solution is stable during storage at room temperature. In terms of the potential of the polymer to retain water, it should be pointed out that with the equal addition amount, the water-retention rate (the amount of water that can be retained in a material) of HPMC is greater than that of MC. HPMC has higher enzyme resistance than MC, and the possibility of enzymatic degradation of HPMC is much less than that of MC. HPMC has higher adhesion than MC. HPMC can combine with water-soluble polymers and then form a uniform solution of higher viscosity, such as polyvinyl alcohol, starch ethers, and vegetable gums [11].

- Carboxymethyl Cellulose (CMC): It is produced by binding of carboxymethyl groups (-CH2-COOH) to some hydroxyl groups present in the glucopyranose or monomers of cellulose backbone [12]. CMC is also known as cellulose gum [13]. The manufacturing of CMC includes 4 major processes: The **isolation** of alpha-cellulose from the water hyacinth and the synthesis by means of the **alkalization** (alkali-catalyzed reaction) of cellulose with chloroacetic acid with the aid of **carboxymethylation** and eventually the **purification** of the CMC itself to cast off undesirable compounds. In other words, it can be expressed that there are two principal reactions in order to turn the cellulose into CMC: alkalization and carboxymethylation [14]. Compared to MC, which is only soluble in cold water, CMC can be dissolved in both hot and cold water, however, its solubility is still poor. CMC and HPMC showing not same water retention rates in such a way that HPMC results in a higher water retention rate when compared to the same amounts [15].

- Croscarmellose Sodium (CCS): CMC is regularly used as its sodium salt named sodium carboxymethyl cellulose (Sodium CMC, SCMC, or NaCMC) to enhance the solubility. NaCMC has a cross-linked polynomial and so it is also recognized as CCS [16]. This polymer is made in two stages, in the first step, crude cellulose is soaked in sodium hydroxide, and then in the second step, NaCMC is formed during the reaction of cellulose with sodium monochloroacetate [17]. Therefore, the main difference between NaCMC and CMC is that NaCMC is easily soluble in hot and cold water, while CMC is poorly soluble in water. NaCMC is white and free-flowing powder that is very slightly soluble in ether. Nary organic solvents [18].

- Ethyl Cellulose (EC): It is produced by means of the reaction between ethyl chloride and alkali cellulose. EC is insoluble in glycerin, propylene glycol, and water, however soluble in different organic solvents [19].

- Hydroxyethyl Cellulose (HEC): It is a hydroxyethyl ether that is produced by soaking the cellulose with sodium hydroxide and then reacting with ethylene oxide. HES indicates high solubility in water over a wide temperature range even at greater than 50 degree Celsius [20].

In terms of chemical formula, repeating structure and substitution group for cellulose derivatives including MC, HPMC, EC, HEC, and NaCMC are altogether shown in **Figure 2**.

- Hydroxypropyl Cellulose (HPC): HPC is a cellulose-based compound in which hydroxyl groups have been hydroxypropylated. It is produced by reacting alkali cellulose with propylene oxide at high pressure and temperature. In comparison with other water-soluble cellulose ethers, HPC is more plastic and relatively hydrophobic due to the high degrees of hydroxypropylation. Because of its chemical structure, it is absolutely soluble in water and also in polar organic solvents such as methanol, ethanol, isopropyl alcohol (IPA), and acetone. HPC solution has lower viscosity and slow hydration or uptake of water compared to HPMC [21]. Based on the degree of substitution, HPC is divided into two types and solubility behavior of each one is different and which are: High substitute hydroxypropyl cellulose (H-HPC) and low substitute hydroxypropyl cellulose (L-HPC) [22].

- Microcrystalline Cellulose (MCC): It is manufactured with the aid of subjecting to a high shear treatment, at increased temperature and pressure, a reaction combination of a cellulose material, an active oxygen compound and water, for a time effective to depolymerize the cellulose material. MCC is an isolated,

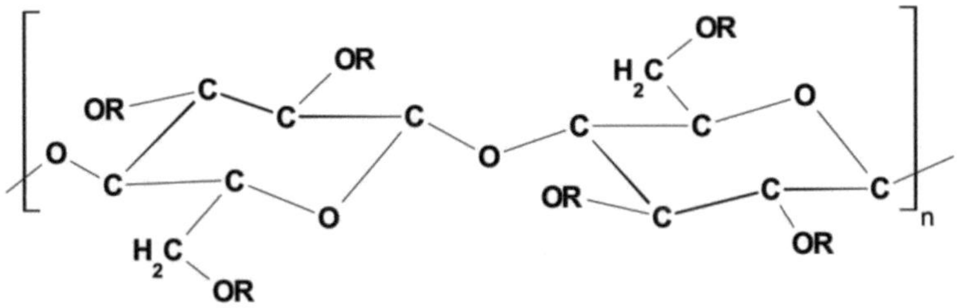

MC	R= H, CH_3
HPMC	R= H, CH_3, $CH_2CH_2CH(OH)\ CH_3$
EC	R= H, CH_2CH_3
HEC	R= H, CH_2CH_2OH
NaCMC	R= H, CH_2COONa

Figure 2.
Repeating structure of cellulose derivatives. R: Substituent group.

colloidal crystalline component of cellulose fibers and an extra purified structure of cellulose. It is a white, odorless, and tasteless carbohydrate polymer powder that usually consists of up to 350 glucose units. It is dispersible in water but not soluble, requiring large strength to disperse and hydrate [23].

- Polyanionic Cellulose (PAC): It is produced from the isopropyl alcohol solution of alkalized cellulose and chloroacetic acid by means of etherification reaction. The raw materials for the obtaining of PAC are comparable to those for the manufacturing of excessive viscosity CMC, however, in the production process, one-of-a-kind degradation technique is employed so that the substitution of hydroxyl group in the ring shape structure of β-glucose group will be extra uniform. PAC shows markedly solubility both in cold water and hot water. It exhibits the best thermal stability, salt-resistance, and strong antibacterial activity. PAC has a molecular aspect similar to CMC [24].

- Hydroxypropyl Methyl Cellulose Phthalate (HPMC-P): It is manufactured using the esterification of HPMC with phthalic anhydride and is also known as hypromellose phthalate. The dissolution properties of HPMC-P can be controlled by altering the phthalyl content that affects the pH ranges solubility. It is soluble in pH 5 to 5.5 [25].

- Hydroxypropyl Methyl Cellulose Acetate Succinate (HPMCAS): It is obtained via the esterification of HPMC with acetic anhydride and succinic anhydride in a reaction medium of a carboxylic acid (acetic acid) and the usage of an alkali carboxylate (sodium acetate) as catalyst. HPMCAS is blended ester of HPMC acetate and succinic acid. HPMCAS begins to swell and dissolve at $pH > 5$ relying on the extent of substitution. The pH of dissolution will increase as the ratio of acetyl over succinoyl substitution becomes greater [26, 27].

- Powdered Cellulose (PC): It is typically produced by means of mechanical micropulverizing cellulose. It has an excessive quantity of amorphous regions and is regularly used as filler and binder in tablet manufacture. In comparison with PC, MCC has a greater degree of crystallinity due to the fact that it is usually obtained via partly hydrolyzing cellulose with mineral acid [28]. It does not dissolve in water, ethanol, ether, and dilute mineral acids and it should be said that PC is slightly soluble in sodium hydroxide solution [29, 30].

- Carboxymethylhydroxyethyl Cellulose (CMHEC): It is a combination of water-soluble cellulose ethers, each with the nature of CMC and HEC. CMHEC has an excessive bonding potential and hydration on the suspended solids with very good flocculation, however, due to nonionic hydroxyl the presence of ethyl group, in contrast with the CMC and salt, it is extra compatible. Also, another advantage is the cross-linking ability of multivalent cation, which can significantly improve the viscosity of solutions [31].

3. Behavior of cellulosic compounds in food systems

A wide range of cellulose derivatives due to their nontoxic nature for a long time have been developed, in particular, acetate, nitrate and sulfate esters, cellulose ethers (MC, HPMC, CMC, EC, HEC, and HPC), and sodium CMC are the most broadly used cellulose derivatives for foodstuffs [32]. **Table 1** indicates the different properties of cellulose-derived compounds, which are used to achieve a specific behavior in food and pharmaceutical systems. As it turns out, the polymers of MC, HPMC, CMC, and HPC exhibit the widest range of properties. Food polymers play an essential role in food structure, food functional properties, food processing, and shelf life. The information in this field is commercially essential as it will provide a beneficial practical guideline for food improvement and industrial production. The application of each of these compounds should be considered in food systems according to the important characteristics of food such as stability, the importance of resistance to stress, maintaining product uniformity, and improving physical properties [36].

Variations of the molecular structure and size of polymer chains or networks consequences in one of kind polymers with special properties. In using each of these polymers, vital behaviors in that food system need to be identified. In a food whose stability in the aqueous phase is important, the preference of a polymer that is soluble in water will be desired. In the design of food formulations based on water-insoluble compounds, it is important to select derivatives with similar properties to the main ingredients of that formulation. Generally, in various applications of cellulose derivatives for the food systems, such as emulsifiers, stabilizers, coatings, film production, gel formation, creating adhesive, and thickening structures, these different and compatible properties of the polymer are considered. Sometimes, several polymers may be suitable for a formulation or purpose, in this situation, a polymer should be chosen that is economical and can be used on an industrial scale. In fact, in a low-scale trial and error, polymers are replaced with each other, and rheological properties and sensory evaluation or textural indicators are tested. In a food system, some positive properties of a polymer may compensate for the disadvantages of another polymer, and a more suitable structure can be created by using a combination. In the research and development departments in the food industry, the application of each of the replaceable polymers is constantly being investigated. Today, the design of stabilizers for different food products has become a profitable industry and business in the world. Many companies supplying industrial raw materials are known for introducing new stabilizing formulations in the food industry [37, 38].

4. Function results of cellulosic structures in foods

Cellulose and its derivatives have shown special roles and results in the industrial production of food and have solved many problems. In the food industry, for a research and development specialist, it is important to know about these roles and their definitions and to substitute different ingredients in a way that is effective. In order to understand the function of cellulose and its derivatives in food systems, its specific roles and applications need to first be defined in following terminology:

- Edible coating (EC): It is described as a thin layer of safe-to-eat material formed on a food. Edible coatings may additionally be described as defensive layers created around food surfaces by using solutions made from suitable for eating polymers like polysaccharides, proteins, lipids, or their combinations. This protective layer acts as a barrier between the food and exterior surroundings and

Properties	**Cellulosic Polymers**												
	MC	**HPMC**	**CMC**	**CCS**	**HPC**	**EC**	**HEC**	**MCC**	**PAC**	**HPMC-P**	**HPMCAS**	**PC**	**CMHEC**
Water solubility	●	●	●	●	●		●		●	●			●
Organic solubility	●	●			●	●				●	●		
Gel forming	●	●	●	●	●	●	●	●	●	●			●
Film forming	●	●	●		●	●	●	●	●	●	●		●
Mucoadhesive system	●	●	●		●	●	●						●
High swelling	●	●	●	●	●		●	●	●	●		●	●
Hydrophilic features	●	●	●	●	●		●	●	●	●		●	●
Hydrophobic features						●					●		
Viscosifying agent	●	●	●	●	●	●	●	●	●	●	●	●	●
Thermoplastic compound								●			●	●	
Drug solubilizer	●	●	●	●	●	●	●	●				●	●

Table 1.
The variety of polymer properties [33–35].

as a result prolongs the ripening and spoilage process. Different kinds of industrial safe-to-eat coatings are broadly used to prevent moisture loss and to add shine to fruits and vegetables. Other business purposes of edible coating consist of coating nuts, fried foods, seafood, minimally processed foods, vegetables, and especially fruits [39].

- Edible Film (EF): It is a preformed, thin layer, made of safe-to-eat material, which once formed can be positioned on or between food components. The main distinction between EC and EF is that the EC is applied in liquid form on the food, commonly through immersing the product in a prepared solution, and EF is first molded as solid sheets using polymers as edible film formers, which are then applied as a wrapping on the food product [40].

- Emulsification: It is the technique of dispersing two or greater immiscible liquids together to structure a semi-stable mixture. In food applications, these two liquids commonly consist of an organic type (oil) section and an aqueous type (water) section that is stabilized by means of the addition of a food-grade emulsifier or surface-active agent. Emulsifiers work by means of forming physical barriers that maintain droplets from coalescing. The surface-active agent or surfactant includes both polar head and nonpolar tail. Therefore, emulsifier as a surfactant is attracted to both hydrophilic and hydrophobic compounds and assists products containing immiscible food ingredients, such as oil and water, to combine [41]. The water-holding capacity of cellulose derivatives such as carboxymethyl cellulose can be used in the formulation of food products such as doughnuts to prevent their staleness by maintaining the product's moisture during storage. Also, in interaction with other oil-uptake inhibitors hydrocolloid polymers, such as xanthan, their negative effects on the texture of the final product were eliminated using CMC and it helps to maintain the proper volume of the final product [42].

- Stabilizing agent: It is an ingredient added to foods to assist, keep, or decorate their unique texture and physical and chemical attributes. Stabilizing agents or stabilizers serve both the practical purpose of maintenance while additionally making products some distance extra appetizing to consumers. They can also work synergistically with emulsifiers to permit food ingredients that would now not in any other case combine well and hold a homogenous dispersion. This will increase the stability and viscosity of the food by means of binding its giant molecules. This component is used for a number of features in foods and its major characteristic is to act as a thickening agent to gel the ingredients into the required consistency [43].

- Rheology control: It is important in the food industry in order to maintain product quality. Rheology of food is tofind out about the consistency and flow of food under certain utilized forces, to recognize the underlying physicochemical concepts of food substances and their interaction. Rheology modifiers assist to obtain preferred rheological behavior, supporting to manage a range of characteristics, consisting of shelf stability, ease of application, sag or bend, resistance, leveling, settling, film forming, regulation compliance, and more [44].

- Suspending agent: It is a substance added to fluids or liquids to help suspend or disperse particles and reduce sedimentation. This is important in the stability of beverages and maintaining the integrity of liquid products that contain fine solids. If the suspension is no longer maintained, particularly in some beverage products, the structure of the product will change in taste, decrease in quality, and not be accepted by the consumer [45].

- Diffusion barrier: It is a thin layer of polymer generally positioned in multilayer films between two different layers or on surface of a layer. Gas diffusion barriers are vital for a number of purposes in food packaging's capability to resist the absorption of light, moisture, and oxygen. These properties come from the accumulation of the individual films used in the development of the packaging. The oxygen, water vapor, and carbon dioxide barrier properties are of specific significance in food packaging applications. The correct barrier and proper gas combination will make the shelf life of food products longer. Several physical factors can have an effect on the barrier properties of a polymer that consist of temperature, humidity, orientation, and cross-linking [46].

- Encapsulation: It is a technique or industry term in which active agents or dietary supplements are surrounded by using a coating to supply small capsules with controlled migration. This procedure includes the incorporation of food ingredients or different substances in small capsules. Also, encapsulation refers to an object's capability to conceal taste and smell that are now not necessary to sense by its consumer [47].

- Extrusion: It is a mechanical procedure in which certain substances as soft blended components are forced, underneath pressure, *via* a die or perforated plate opening to create products of a preferred shape, structure, size, and/or texture. This technique is used in food processing and described as a machine of pushing out, combined components in form of paste, especially *via* a small opening referred to as a die that is designed to produce the required shape [48].

- Molding: The operation of molding is to form the dough piece, in accordance to the bread variety being produced, so that it appropriate fits and matches into pans. The dough molding tools can be set to obtain the preferred form with a minimal quantity of stress and strain on the dough [49].

- Foam stabilizer: Liquid foams are often made pretty long-lasting by means of including some substance, known as a foam stabilizer, that prevents or retards the coalescence of the gas bubbles. A stronger foam structure is to add polymer to a foam-making system. The polymer will increase the liquid viscosity for that reason extend the foam stability and also additionally minimize the adsorption of foaming agents. Thus, the synergistic effect in a foam system is achieved [50].

- Thermoplastic system: It is thermo-softening plastic or binder that turns soft and flexible at certain temperatures and solidifies when cooled [51].

Table 2 shows the main specific use of cellulose derivatives in the food and drug industry.

Application	Cellulosic compound	Reference
Edible coating	MC, HPMC, and NaCMC	[52, 53]
Edible Film	MC, HPMC, CMC, andHPC	
Emulsification	EC and NaCMC	[54, 55]
Stabilizing agent	EC, HPC, and NaCMC	
Rheology control	MC, HPMC, NaCMC, and HPC	[56]
Suspending agent	MC, HPMC, NaCMC, and HPC	
Diffusion barrier	MCC	[57, 58]
Encapsulation	HEC, HPC, HPMC, MC, and NaCMC	[59]
Extrusion	HPMC	[60]
Molding	HPMC	
Foam stabilizer	MC, HPMC, NaCMC, andMCC	[61]
Thermoplastic system	CMC and MCC	[62]

Table 2.
The main applications of cellulose compounds.

5. Food safety of cellulosic compounds

Reevaluation of the food safety of cellulosic compounds has been carried out by international associations such as the European Food Safety Authority (EFSA) in the EFSA's scientific panels on Food Additives and Nutrient Sources added to Food (ANS). Food safety of cellulose derivatives as food additives including MCC (E 460(i)), PC (E 460(ii)), MC (E 461), EC (E 462), HPC (E 463), HPMC (E 464), ethyl MC (E 465), NaCMC (E 466), enzymatically hydrolyzed CMC (E 469), and cross-linked CMC (E 468) has been investigated ("E numbers" are special codes for components used as food additives and "E" is the abbreviation of Europe) [63]. The ideal everyday consumption as an acceptable daily intake (ADI) "not distinguished" for unmodified and modified celluloses established by the Joint FAO/WHO Expert Committee on Food Additives (JECFA) and the Scientific Committee on Food (SCF). Celluloses are not absorbed in the human body and are excreted intact through the feces after digestion of foods. Additionally, MCC, PC and modified celluloses are fermentable by means of intestinal flora in animals and humans. Specific toxicity facts are not available for all different celluloses, which were considered common among all types of cellulose due to their structural and biological similarities. In general, the acute toxicity of all cellulose types was low and there was no genotoxic concern. Short-term and sub-chronic dietary toxicity studies conducted with MCC, Mc, EC, HPC, HPMC, NaCMC, and enzymatically hydrolyzed CMC showed no specific treatment-related adverse symptoms and unfavorable side effects at levels up to 10%. In the performed studies about chronic toxicity with MCC, MC, HPC, HPMC, ethyl MC, and NaCMC, there is no side effect reported up to 9000 mg/kg body weight per day. The carcinogenicity effects were not observed for MCC and other cellulose derivatives. The reproductive problems and growth side effects were not indicated with doses of cellulose higher than 1000 mg/kg. The daily exposure to a variety of cellulosic compounds such as MCC, NaCMC, cross-linked CMC, and enzymatically hydrolyzed CMC for the general population was up to 506 mg/kg body weight. It was reported that there is no requirement for a numerical ADI and there would be no safety concern in consumption and use levels for all considered types of cellulose. Finally, the panel reported the daily indicative total exposure of cellulose derivatives around 660–900 milligrams per kilogram of human body weight. The safety of cellulose and cellulose derivatives has been extensively studied in animals and humans. These studies show that cellulose and cellulose derivatives pass through the gastrointestinal tract unchanged and can be rapidly detected in the feces of experimental animals and humans upon consumption. JECFA stated that viscosity is not recognized as a factor related to the safety of these additives. Many safety research reviewed and analyzed the use of cellulose compounds. All of these research aid the statement that there is no safety impact arising from a variation in viscosity. In general, if it is not definitely claimed that cellulose is nontoxic, it can at least be said that celluloses are of very low toxicity at the ranges of consumption and intake imperative to obtain the preferred impact and do now not pose a hazard or risk to health. Cellulose and its derivatives both are known to be safe in terms of environmental impacts [64, 65].

6. The principles of food regulation in the use of cellulose derivatives

The Code of Federal Regulations (CFR) is an arrangement and codification of general, customary, and everlasting regulations published in the Federal Register by federal government agencies, supervisory departments, and businesses. In relation to the legal guidelines related to the consumption of special raw materials in the food industry, considerable codes have been provided that present appropriate information to industrial producers [66]. In the case of drugs for human use, such as ophthalmic demulcents, the concentrations percentage of each cellulose derivative (NaCMC, HEC, HPMC-P, and MC) has been established within 0.2 to 2.5% in accordance with the standard of 21CFR349.12 [67]. The use of ethyl cellulose in food and feed is stated in the legal standards of 21CFR172.868 and 21CFR573.420, respectively. Ethyl cellulose as food additive may be safely used in foods under the stated conditions: In clause “a”—The food additive is a cellulose ether containing ethoxy groups (OC2H5) linked by an ether bond and containing not more than 2.6 ethoxy groups per anhydro glucose unit on an anhydrous basis. In clause “b”—Ethyl cellulose is used or intended for use as a binder and filler in dry vitamin preparations, as part of protective coatings for vitamin and mineral tablets, and finally in the role of stabilizer in flavoring compounds. For the use of ethyl cellulose as animal feed, it must meet the previously stated conditions of clause “a” in the same way as in clause “b–”—its uses are presented as a binder or filler in dry vitamin preparations to be incorporated into animal feed [68, 69]. Usage rules for hydroxypropyl cellulose are provided in 21CFR172.870. This standard separates HPC with a high degree of substitution (which contains no more than 4.6 hydroxypropyl groups per anhydro glucose unit) and a low degree of substitution (which contains on average 0.1 to 0.4 hydroxypropyl groups per anhydro glucose unit). High substituted HPC may be used as an emulsifier, film former, protective colloid, stabilizer, suspending agent, and thickener according to good manufacturing practice (GMP). HPC with low substitution can be used as a binder and disintegrator in tablets or wafers containing dietary supplements according to GMP. FDA indicated that the proposed use of high-substituted HPC is safe with a minimum viscosity of 10 centipoises and for same using as a binder in dietary supplements similar to low-substituted HPC will not result in an increased intake or harm to human health under the established conditions of use and significantly it is not expected to have different biological properties [70]. Finally, for HPC based on federal regulation of 75FR17928, there is no significant effect on the environment and thus an environmental impact statement is not required [71]. The specific labeling requirements for the use of various hydrophilic gums such as NaCMC in specific drug products are noted in 21CFR201.319 [72].

The prescribed conditions for the safe use of methyl ethyl cellulose are provided in the standard of 21CFR172.872: In clause "a"—The general formula of cellulose ether is as $[C_6H_{(10\text{-}x\text{-}y)}\ O_5(CH_3)_x(C_2H_5)_y]_{n.}$ In this chemical formula, x is the number of methyl groups and y is the number of ethyl groups. The average value of x is 0.3 and the average value of y is 0.7. In clause "b"—This additive composition must have the following conditions: The methoxy content shall be not less than 3.5% and not more than 6.5%, calculated as OCH_3, and the ethoxy content shall be not less than 14.5% and not more than 19%, calculated as OC_2H_5, both measured on the dry sample. The viscosity of an aqueous solution, 2.5 grams of the material in 100 milliliters of water, at 20°C, is 20 to 60 centipoises. The ash content on a dry basis has a maximum of 0.6%. In clause "c"—The food additive is used as an aerating, emulsifying, and foaming agent, in an amount not in excess of that reasonably required to produce its intended effect [73].

The specified conditions for safely using "Hydroxyethyl cellulose film-water insoluble" for packaging food are stated in accordance with the regulation code of 21CFR177.1400. In clause "a"—Water-insoluble hydroxyethyl cellulose film consists of a base sheet manufactured by the ethoxylation of cellulose under controlled conditions, to which may be added certain optional substances of a grade of purity suitable for use in food packaging as constituents of the base sheet or as coatings applied to impart desired technological properties. In clause "b"—The optional substances that are identified as suitable in terms of food safety can be used in the base sheet of film production and coating as components of water-insoluble hydroxyethyl cellulose film. In clause "c"—Any substance employed in the production of the water-insoluble hydroxyethyl cellulose film described in this section [74].

7. Innovations of the cellulosic application in the food industry

Studies discuss the interrelation between the chemical structure of cellulose and its source and its various physicochemical characteristics. Although cellulose extracted from plants has been most investigated, cellulose purified from microorganisms and animals with special structural features has received increasing attention [75].

Heretofore, enzymatic hydrolysis procedure, TEMPO oxidation reaction (2,2,6,6-Tetramethylpiperidine-1-oxyl), and carboxymethylation process have been widely used to assist in defibrillating cellulose during processing into nanocellulose, which not only helps decrease consumption of energy but also provides additional functional groups for the final products [76].

Acidic hydrolysis, which includes inorganic and organic acids, can remove and eliminate amorphous regions, resulting in cellulose nanocrystal (CNC), although highly corrosive conditions and low CNC yield are persistent issues. Through mechanical treatments such as refining, homogenization, microfluidization, sonication, ball milling, and the aqueous counter collision (ACC) technique, cellulose nanofibrils (CNF) can be produced, however, the very high input energy prohibits the commercialization of these methods [77, 78].

Bacterial cellulose (BC), a special nanocellulose derived from bacteria, has currently attracted numerous research interests. Its higher aspect ratio and larger diameter make it a promising material for food packaging applications. In order to facilitate the application of nanocellulose in food packaging, it is always processed using different manufacturing technologies, including solution casting, layer-by-layer (LBL) assembly, extrusion, coating, gel-forming, spray drying, electrostatic spinning, adsorption, with the aim of producing different forms of materials such as film, gel, coating membrane, and emulsions [79, 80].

Due to their non-toxicity, proper biodegradability and biocompatibility, high aspect ratio, low coefficient of thermal expansion, extremely good mechanical strength, and special optical properties, nanocellulose-based food packaging materials are broadly used for fruit packaging, meat products, fast foods, dairy products, and beverages. Since nanocellulose and functional fillers in cellulose-based nanocomposites provide excellent barrier and mechanical properties, antibacterial activity and stimuli-responsive performance, therefore noticeably improve food quality stability and shelf life [81, 82].

Nanocellulose and its derived nanocomposite have turned out to be a hotspot of research in food packaging due to its great properties including excessive strength, large specific surface area, excellent barrier properties and proper biocompatibility, safety, non-toxicity, and degradability. In food packaging materials, nanocellulose-based nanocomposites can be used as fresh and antibacterial packaging materials, smart packaging materials, and high-barrier packaging materials, which indicates the excessive utility potential of nanocellulose-based composites. Therefore, nanocellulose-based composites, as a type of renewable and environmentally friendly packaging material, enhance the protection and quality of food and are one of the important paths to understand the improvement of "eco-friendly industry" in the food industry [83].

Recently, biopolymer-based hydrogels have been attracting growing attention due to the fact of their promising applications in drug release, biosensors, superabsorbent materials, tissue engineering, and many others [84]. Hydrogels are water-insoluble polymers that have the ability to hold a large amount of water in their network [85, 86]. The considerable hydrophilic groups with hydroxyl, carboxyl, and aldehyde form in the structure of native cellulose or cellulose derivatives make them suitable to prepare hydrogels without difficulty with fascinating constructions and properties for biomedical and food packaging and applications [87, 88].

Regarding the progress of cellulose and its derivatives in the food industry, especially packaging, the following should be mentioned [89, 90]:

- Expanding cellulose resources to different biomass besides traditional raw materials, such as tunicate and BC, for higher quality and greater satisfactory nanocellulose to develop and improve many advanced applications. So far, commercially available nanocellulose is mainly produced from wood and other plant sources. Although numerous research works focused on the preparation of nanocellulose from tunicate and BC, confirmed their higher overall performance than wood-based nanocellulose, further research and evaluation on the preparation-properties-performance correlation of this particular nanocellulose is important and necessary.

- Developing an efficient, cost-effective, eco-friendly, or environmentally-friendly approach for extraction of nanocellulose. Although several new extraction methods have been developed, the most broadly used ones are sulfuric acid hydrolysis and mechanical refining. However, the harsh and extreme acid hydrolysis, high water usage, large amount of polluted wastewater, high demand for energy, and low efficiency have largely prohibited the industrially feasible production of nanocellulose. Therefore, more work efforts should be made to develop novel methods for the preparation of nanocellulose, such as those based on organic acids, which have already been shown to be a green approach (not environmentally harmful strategy) for the preparation of functionalized derivatives as nanocellulose.

- Elaborating new procedures to fabricate cellulosic nanocomposites. In the laboratory, the solution casting technique is still broadly used for research purposes, which is not suitable for industrial-scale production. In the pilot scale, extrusion has been used, although it is not an excellent approach due to the fact the nanocellulose is evermore dispersed in water, which negatively impacts the performance of the extrusion. Accordingly, it is essential to develop a scalable method for the preparation of nanocellulose-based composites for food packaging materials.

- Improving the overall performance of nanocellulose-based composites as packaging materials. The perfect food packaging materials require anti-UV (UV-proof, UV-block, or UV-protect), vapor and gas barrier properties, excellent mechanical strength, and appropriate hydrophobicity. Particularly for the last one, new techniques have to be developed to change the hygroscopic nature of nanocellulose and increase the wet strength or wet tensile value, hence making its functions and applications in daily life more practical. For example, the coating with natural wax or esterification reaction as a pretreatment seems appropriate to achieve this goal.

- Development of smart packaging materials based on nanocellulose. At present, achieving the responsive properties of cellulose nanocomposites is usually performed with the aid of combining different organic and inorganic fillers. However, the migration and release of functional fillers and their potential health risks have not been comprehensively investigated. Future researchers in systematic studies should not only pay attention to the safety issue of nanocellulose itself but also focus on the functionality of selected fillers.

- Designing packaging materials based on nanocellulose for food. Although many studies generally focused on the preparation and properties of packaging materials, little attention is paid to the interaction between materials and food and even neglected the various aspects that affect the application of materials. For example, the impact of environmental prerequisites and conditions on the quality change of both food products and packaging materials has to be investigated to show the feasibility and suitability of packaging materials for a particular food.

8. Conclusions and outlook

In summary, cellulose has been used in various food applications fields for decades, and its impact on the food industries has become increasingly important. The porous network, biocompatibility, high aspect ratio, and abundance of hydroxyl groups are advantages of cellulose for use as a functional food component. Cellulose creates different colloidal conditions such as surface charge, dimensions, crystal structure, solubility, and wettability. The colloidal properties of cellulosic polymers can be controlled by changing the manufacturing parameters or replacing surface hydroxyl groups with the ester groups. In addition, health benefits and functional impacts such as hypoglycemic and hypolipidemic activity and intestinal bioavailability due to cellulose can be affected by colloidal states. Based on the review of major studies related to cellulose in the food industry, it is possible to refer to the following cases for valuable consideration as cellulose-related challenges in the future: (1) the design of functional foods is possible by changing the colloidal states of cellulose, (2) new food properties can be obtained by different colloidal states in food matrix (3) the interaction between cellulose and gut microbiota needs to be further studied and the potential mechanism needs further investigation, (4) the application of nanocellulose in food matrix is increasingly discovered and the potential long-term risk of nanocellulose requires regulatory and safety testing, and (5) the energy consumption and potential environmental pollution should be considered, and new green technologies to extract or prepare different types of cellulose are proposed.

Conflict of interest

The author declares no conflict of interest.

Notes

This chapter was written in such a way that those whose specialty is not food science and technology can understand the uses and effects of cellulose and its derivatives in the food industry.

Author details

Hassan Sabbaghi
Ferdowsi University of Mashhad, Mashhad, Iran

*Address all correspondence to: hassansabbaghi@gmail.com

References

[1] Ullah H, Santos HA, Khan T. Applications of bacterial cellulose in food, cosmetics and drug delivery. Cellulose. 2016;**23**(4):2291-2314

[2] Ateş S, Durmaz E, Hamad A. Evaluation possibilities of cellulose derivatives in food products. Kastamonu University Journal of Forestry Faculty. 2016;**16**(2):383-400

[3] Sun S, Sun S, Cao X, Sun R. The role of pretreatment in improving the enzymatic hydrolysis of lignocellulosic materials. Bioresource Technology. 2016;**199**:49-58

[4] Dhingra D, Michael M, Rajput H, Patil RT. Dietary fibre in foods: A review. Journal of food science and Technology. 2012;**49**(3):255-266

[5] Coffey DG, Bell DA, Henderson A. Cellulose and cellulose derivatives. Food Polysaccharides and Their Applications. 2006;**2**:146-179

[6] Lavanya DK, Kulkarni PK, Dixit M, Raavi PK, Krishna LN. Sources of cellulose and their applications—A review. International Journal of Drug Formulation and Research. 2011;**2**(6):19-38

[7] Siqueira G, Bras J, Dufresne A. Cellulosic bionanocomposites: A review of preparation, properties and applications. Polymers. 2010;**2**(4):728-765

[8] Vieira JG, Rodrigues Filho G, Meireles CD, Faria FA, Gomide DD, Pasquini D, et al. Synthesis and characterization of methylcellulose from cellulose extracted from mango seeds for use as a mortar additive. Polímeros. 2012;**22**:80-87

[9] Greiderer A, Steeneken L, Aalbers T, Vivó-Truyols G, Schoenmakers P. Characterization of hydroxypropylmethylcellulose (HPMC) using comprehensive two-dimensional liquid chromatography. Journal of Chromatography A. 2011;**1218**(34):5787-5793

[10] Garner J, Park K. Chemically modified natural polysaccharides to form gels. In: Ramawat KG, Mérillon JM, editors. Polysaccharides. Champions. Switzerland: Springer; 2015. pp. 1555-1582

[11] Grover JA. Methylcellulose (MC) and hydroxypropylmethylcellulose (HPMC). In: Food Hydrocolloids. Netherlands: CRC Press; 2020. pp. 121-154

[12] Yang GM, Fan XH, Chen XL, Yuan LS, Huang XX, Li X. Interaction mechanism between carboxylmethyl cellulose and iron ore concentrates in iron ore agglomeration. Journal of Central South University. 2015;**22**(4):1241-1246

[13] Prenzo A. The world of cellulose gums. Food Product Design. 2011;**21**(5):1-3

[14] Saputra AH, Qadhayna L, Pitaloka AB. Synthesis and characterization of carboxymethyl cellulose (CMC) from water hyacinth using ethanol-isobutyl alcohol mixture as the solvents. International Journal of Chemical Engineering and Applications. 2014;**5**(1):36

[15] Mallikarjunan P, Chinnan MS, Balasubramaniam VM, Phillips RD. Edible coatings for deep-fat frying of starchy products. LWT-Food Science and Technology. 1997;**30**(7):709-714

[16] Lopez CG, Rogers SE, Colby RH, Graham P, Cabral JT. Structure of

sodium carboxymethyl cellulose aqueous solutions: A SANS and rheology study. Journal of Polymer Science Part B: Polymer Physics. 2015;**53**(7):492-501

[17] Gilbert M. Cellulose plastics. In: Gilbert M, editor. Brydson's Plastics Materials. Amsterdam: Butterworth-Heinemann; 2017. pp. 617-630

[18] Hiremath P, Nuguru K, Agrahari V. Material attributes and their impact on wet granulation process performance. In: Narang A, Badawy S, editors. Handbook of Pharmaceutical Wet Granulation. United States of America: Academic Press; 2019. pp. 263-315

[19] Wasilewska K, Winnicka K. Ethylcellulose–A pharmaceutical excipient with multidirectional application in drug dosage forms development. Materials. 2019;**12**(20):3386

[20] Arai K, Shikata T. Hydration/dehydration behavior of hydroxyethyl cellulose ether in aqueous solution. Molecules. 2020;**25**(20):4726

[21] Klug ED. Some properties of water-soluble hydroxyalkyl celluloses and their derivatives. Journal of Polymer Science Part C: Polymer Symposia. 1971;**36**(1):491-508

[22] Mishra SM, Sauer A. Effect of physical properties and chemical substitution of excipient on compaction and disintegration behavior of tablet: A case study of low-substituted hydroxypropyl cellulose (L-HPC). Macromolecules. 2022;**2**(1):113-130

[23] Thoorens G, Krier F, Leclercq B, Carlin B, Evrard B. Microcrystalline cellulose, a direct compression binder in a quality by design environment—A review. International Journal of Pharmaceutics. 2014;**473**(1-2):64-72

[24] Vasconcelos NF, Feitosa JP, Andrade FK, Miranda MA, Sasaki JM, Morais JP. Chemically modified cellulose nanocrystals as polyanion for preparation of polyelectrolyte complex. et al., Cellulose. 2019;**26**(3):1725-1746

[25] Shi SC. Hydroxypropyl methylcellulose phthalate biopolymer as an anticorrosion coating. International Journal of Electrochemical Science. 2021;**16**(9):1-10

[26] Miller WK, Lyon DK, Friesen DT, Caldwell WB, Vodak DT, Dobry DE, et al. Hydroxypropyl methyl cellulose acetate succinate with enhanced acetate and succinate substitution. United States patent US 9,040,033. 26 May 2015

[27] Wang S, Liu C, Chen Y, Zhu A, Qian F. Aggregation of hydroxypropyl methylcellulose acetate succinate under its dissolving pH and the impact on drug supersaturation. Molecular Pharmaceutics. 2018;**15**(10):4643-4653

[28] Fechner PM, Wartewig S, Füting M, Heilmann A, Neubert RH, Kleinebudde P. Properties of microcrystalline cellulose and powder cellulose after extrusion/spheronization as studied by Fourier transform Raman spectroscopy and environmental scanning electron microscopy. AAPS PharmSci. 2003;**5**(4):77-89

[29] Lenhart V, Quodbach J, Kleinebudde P. Mechanistic understanding regarding the functionality of microcrystalline cellulose and powdered cellulose as pelletization aids in wet-extrusion/spheronization. Cellulose. 2020;**27**(4):2189-2210

[30] Isogai A, Atalla RH. Dissolution of cellulose in aqueous NaOH solutions. Cellulose. 1998;**5**(4):309-319

[31] Sun R, Fang B, Lu Y, Qiu X, Du W, Han X, et al. Rheological properties of hexadecyl dimethyl amine modified

carboxymethyl hydroxyethyl cellulose solutions and its gelling process. Journal of Dispersion Science and Technology. 2018;**39**(1):138-142

[32] Richardson S, Gorton L. Characterisation of the substituent distribution in starch and cellulose derivatives. Analytica Chimica Acta. 2003;**497**(1-2):27-65

[33] Harika K, Sunitha K, Kumar PP, Maheshwar K, Rao MY. Basic concepts of cellulose polymers-a comprehensive review. Archives of Pharmacy Practice. 2012;**3**(3):202

[34] Kamel S, Ali N, Jahangir K, Shah SM, El-Gendy AA. Pharmaceutical significance of cellulose: A review. Express Polymer Letters. 2008;**2**(11):758-778

[35] Sun B, Zhang M, Shen J, He Z, Fatehi P, Ni Y. Applications of cellulose-based materials in sustained drug delivery systems. Current Medicinal Chemistry. 2019;**26**(14):2485-2501

[36] Doelker E. Cellulose derivatives. In: Biopolymers I. Berlin, Heidelberg: Springer; 1993. pp. 199-265

[37] Jayakody MM, Kaushani KG, Vanniarachchy MP, Wijesekara I. Hydrocolloid and water soluble polymers used in the food industry and their functional properties: A review. Polymer Bulletin. 2022;**79**:1-26

[38] Lapasin R, Pricl S. Industrial applications of polysaccharides. In: Rheology of Industrial Polysaccharides: Theory and Applications. Boston, MA: Springer; 1995. pp. 134-161

[39] Zaritzky N. Edible coatings to improve food quality and safety. In: Food Engineering Interfaces. New York, NY: Springer; 2010. pp. 631-659

[40] Falguera V, Quintero JP, Jiménez A, Muñoz JA, Ibarz A. Edible films and coatings: Structures, active functions and trends in their use. Trends in Food Science & Technology. 2011;**22**(6):292-303

[41] Krog N. Food emulsifiers. In: Lipid Technologies and Applications. United States of America: Routledge; 2018. pp. 521-534

[42] Sabbaghi H. Application of hydrocolloid compounds (xanthan and carboxymethylcellulose) in doughnut formulation for reducing oil uptake. Iranian Food Science and Technology Research Journal. 2021;**17**(5):919-940

[43] Himashree P, Sengar AS, Sunil CK. Food thickening agents: Sources, chemistry, properties and applications-A review. International Journal of Gastronomy and Food Science. 2022;**27**:100468

[44] Braun DD, Rosen MR. Rheology Modifiers Handbook: Practical Use and Application. United States of America: Elsevier; 2013

[45] Jindal N, Khattar JS. Microbial polysaccharides in food industry. In: Grumezescu A, Holban AM, editors. Biopolymers for Food Design. United States of America: Academic Press; 2018. pp. 95-123

[46] Scarfato P, Di Maio L, Incarnato L. Recent advances and migration issues in biodegradable polymers from renewable sources for food packaging. Journal of Applied Polymer Science. 2015;**132**(48):1-11

[47] Nedović V, Kalušević A, Manojlović V, Petrović T, Bugarski B. Encapsulation systems in the food industry. In: Advances in Food Process Engineering Research and Applications. Boston, MA: Springer; 2013. pp. 229-253

[48] Wolf B. Polysaccharide functionality through extrusion processing. Current Opinion in Colloid & Interface Science. 2010;**15**(1-2):50-54

[49] Lipton JI, Cutler M, Nigl F, Cohen D, Lipson H. Additive manufacturing for the food industry. Trends in food Science & Technology. 2015;**43**(1):114-123

[50] Murray BS. Recent developments in food foams. Current Opinion in Colloid & Interface Science. 2020;**50**:101394

[51] Sharma S, Singh J, Kumar H, Sharma A, Aggarwal V, Gill AS, et al. Utilization of rapid prototyping technology for the fabrication of an orthopedic shoe inserts for foot pain reprieve using thermo-softening viscoelastic polymers: A novel experimental approach. Measurement and Control. 2020;**53**(3-4):519-530

[52] Zhang Y, Rempel C, Mclaren D. Edible coating and film materials: Carbohydrates. In: Han JH, editor. Innovations in Food Packaging. United States of America: Academic Press; 2014:305-323

[53] Osorno DM, Castro C. Cellulose application in food industry: A review. In: Somashekar R, Thejas Urs G, editors. Emergent Research on Polymeric and Composite Materials. United States of America: IGI Global; 2018:38-77

[54] Hon DN. Cellulose and its derivatives: Structures, reactions, and medical uses. In: Polysaccharides in Medicinal Applications. United States of America: Routledge; 2017. pp. 87-105

[55] Wüstenberg T. Cellulose and Cellulose Derivatives in the Food Industry: Fundamentals and Applications. Germany: John Wiley & Sons; 2014

[56] Zheng YJ, Loh XJ. Natural rheological modifiers for personal care. Polymers for Advanced Technologies. 2016;**27**(12):1664-1679

[57] Bilbao-Sainz C, Avena-Bustillos RJ, Wood DF, Williams TG, McHugh TH. Composite edible films based on hydroxypropyl methylcellulose reinforced with microcrystalline cellulose nanoparticles. Journal of Agricultural and Food Chemistry. 2010;**58**(6):3753-3760

[58] Miller KS, Krochta JM. Oxygen and aroma barrier properties of edible films: A review. Trends in Food Science & Technology. 1997;**8**(7):228-237

[59] Shah H, Jain A, Laghate G, Prabhudesai D. Pharmaceutical excipients. In: Remington. United States of America: Academic Press; 2021. pp. 633-643

[60] Matsuoka T, Kumagai K, Nonaka H. Thermal extrusion of cellulose using hydroxypropyl methylcellulose. Cellulose. 2022;**29**(5):2975-2983

[61] Kumar A, Singh K, Gupta SP. Application of cellulose derivatives in mineral processing. In: Cellulose Science and Derivatives. London, UK, London: IntechOpen; 2021

[62] Stanley WF, Bandaru AK, Rana S, Parveen S, Pichandi S. Mechanical, dynamic-mechanical and wear performance of novel non-crimp glass fabric-reinforced liquid thermoplastic composites filled with cellulose microcrystals. Materials & Design. 2021;**212**:110276

[63] Shatkin JA, Kim B. Environmental Health and safety of cellulose nanomaterials and composites. Handbook of Nanocellulose and Cellulose Nanocomposites. 2017;**2**:683-729

[64] EFSA Panel on Food Additives and Nutrient Sources added to Food (ANS),

Younes M, Aggett P, Aguilar F, Crebelli R, Di Domenico A, et al. Re-evaluation of celluloses E 460 (i), E 460 (ii), E 461, E 462, E 463, E 464, E 465, E 466, E 468 and E 469 as food additives. EFSA Journal. 2018;**16**(1):e05047

[65] EFSA Panel on Food Additives and Nutrient Sources added to Food (ANS). Safety of a change in specifications for the food additive hydroxypropyl methyl cellulose (E 464). EFSA Journal. 2015;**13**(5):4088

[66] McKinney RJ. A Research Guide to the Federal Register and the Code of Federal Regulations. United States of America: Law Librarians' Society of Washington, DC; 2006

[67] Food and Drug Administration. 21CFR349. 12. Available from: https://www.accessdata.fda.gov/scripts/cdrh/cfdocs/cfcfr/CFRSearch.cfm?fr=349.12

[68] Food and Drug Administration. 21CFR172.868. Available from: https://www.accessdata.fda.gov/scripts/cdrh/cfdocs/cfcfr/CFRSearch.cfm?fr=172.868

[69] Food and Drug Administration. 21CFR573.420. Available from: https://www.accessdata.fda.gov/scripts/cdrh/cfdocs/cfcfr/cfrsearch.cfm?fr=573.420

[70] Food and Drug Administration. 21CFR172.870. Available from: https://www.accessdata.fda.gov/scripts/cdrh/cfdocs/cfcfr/CFRSearch.cfm?fr=172.870

[71] Federal Register. 75FR17928. Available from: https://www.federalregister.gov/citation/75-FR-17928

[72] Food and Drug Administration. 21CFR201.319. Available from: https://www.accessdata.fda.gov/scripts/cdrh/cfdocs/cfcfr/CFRSearch.cfm?fr=201.319

[73] Food and Drug Administration. 21CFR172.872. Available from: https://www.accessdata.fda.gov/scripts/cdrh/cfdocs/cfcfr/CFRSearch.cfm?fr=172.872

[74] Food and Drug Administration. 21CFR177.1400. Available from: https://www.accessdata.fda.gov/scripts/cdrh/cfdocs/cfcfr/cfrsearch.cfm?fr=177.1400

[75] Trache D, Hussin MH, Haafiz MM, Thakur VK. Recent progress in cellulose nanocrystals: Sources and production. Nanoscale. 2017;**9**(5):1763-1786

[76] Li Z, Zhang Y, Anankanbil S, Guo Z. Applications of nanocellulosic products in food: Manufacturing processes, structural features and multifaceted functionalities. Trends in Food Science & Technology. 2021;**113**:277-300

[77] Nechyporchuk O, Belgacem MN, Bras J. Production of cellulose nanofibrils: A review of recent advances. Industrial Crops and Products. 2016;**93**:2-5

[78] Zhai L, Kim HC, Kim JW, Choi ES, Kim J. Cellulose nanofibers isolated by TEMPO-oxidation and aqueous counter collision methods. Carbohydrate Polymers. 2018;**191**:65-70

[79] Blanco A, Monte MC, Campano C, Balea A, Merayo N, Negro C. Nanocellulose for industrial use: Cellulose nanofibers (CNF), cellulose nanocrystals (CNC), and bacterial cellulose (BC). In: Handbook of Nanomaterials for Industrial Applications. United States of America: Elsevier; 2018. pp. 74-126

[80] Huang Y, Zhu C, Yang J, Nie Y, Chen C, Sun D. Recent advances in bacterial cellulose. Cellulose. 2014;**21**(1):1-30

[81] Dufresne A. Nanocellulose processing properties and potential applications. Current Forestry Reports. 2019;**5**(2):76-89

[82] Zinge C, Kandasubramanian B. Nanocellulose based biodegradable polymers. European Polymer Journal. 2020;**133**:109758

[83] Wang J, Han X, Zhang C, Liu K, Duan G. Source of nanocellulose and its application in nanocomposite packaging material: A review. Nanomaterials. 2022;**12**(18):3158

[84] Dong S, Feng S, Liu F, Li R, Li W, Liu F, et al. Factors influencing the adhesive behavior of carboxymethyl cellulose-based hydrogel for food applications. International Journal of Biological Macromolecules. 2021;**179**:398-406

[85] Li J, Jia X, Yin L. Hydrogel: Diversity of structures and applications in food science. Food Reviews International. 2021;**37**(3):313-372

[86] Chang C, Zhang L. Cellulose-based hydrogels: Present status and application prospects. Carbohydrate polymers. 2011;**84**(1):40-53

[87] Fu LH, Qi C, Ma MG, Wan P. Multifunctional cellulose-based hydrogels for biomedical applications. Journal of materials chemistry B. 2019;**7**(10):1541-1562

[88] Batista RA, Espitia PJ, Quintans JD, Freitas MM, Cerqueira MÂ, Teixeira JA, et al. Hydrogel as an alternative structure for food packaging systems. Carbohydrate Polymers. 2019;**205**:106-116

[89] Li J, Zhang F, Zhong Y, Zhao Y, Gao P, Tian F, et al. Emerging food packaging applications of cellulose nanocomposites: A review. Polymers. 2022;**14**(19):4025

[90] Huang S, Liu X, Chang C, Wang Y. Recent developments and prospective food-related applications of cellulose nanocrystals: A review. Cellulose. 2020;**27**(6):2991-3011